This is Our Brain

Jeroen Hendrikse

This is Our Brain

 Springer

Jeroen Hendrikse
Section Neuroradiology,
  Department of Radiology
UMC Utrecht
Utrecht
The Netherlands

ISBN 978-981-13-5062-7   ISBN 978-981-10-4148-8   (eBook)
DOI 10.1007/978-981-10-4148-8

Printed on acid-free paper

This Springer imprint is published by Springer Nature
The registered company is Springer Nature Singapore Pte Ltd.
The registered company address is: 152 Beach Road, #21-01/04 Gateway East, Singapore 189721, Singapore

For Inga and Olivier

Preface

This book was written for everyone who would like to know how our brain really looks like, in real MRI and CT scans rather than schematic representations of our brain. The brain is the part of our body which is most often subjected to medical imaging. Many people now have experience of lying in an MRI or CT scanner. Nearly everyone knows someone—be it a friend or a relative—who has undergone an MRI or CT procedure. In addition to the brain, this book will also discuss the body parts adjacent to the brain, such as the skull, the paranasal sinuses and the vertebral column. This book will both describe and visualise the main abnormalities which can be identified in MRI or CT images. Generally, these will be ageing-related abnormalities and common syndromes such as cerebral infarctions, skull fractures and spinal hernias. In addition to clear descriptions, each subject will be illustrated with an MRI or CT image. Each MRI or CT image will come with a clear description next to it.

Although many things can go wrong with our brain, this book will also show you how strong our brain is. Our brain is capable of overcoming adversity, and often emerge all the stronger because of it.

Utrecht, The Netherlands   Jeroen Hendrikse

The original version of this book was revised. An erratum to this chapter can be found at DOI 10.1007/978-981-10-4148-8_41

Contents

Chapter 1
CT and MRI Scans: The Basic Principles

CT and MRI scanners enable doctors to reduce the human body to slices just a few millimetres thick. Until the arrival of CT and MRI scanners in around 1980, it was not possible to do so. At the time, X-ray images were unable to show us individual slices of the body. To understand why this was important, imagine an unsliced loaf of bread. Unsliced bread does not reveal its inner secrets, much like the skull does not reveal the secrets of the brain. Until 1980, doctors had to guess as to what went on inside a patient's skull. Generally, they were unable to take a detailed look at a patient's brain until the patient had died and they had opened his or her skull. They would then literally slice the dead patient's brain to determine the cause of his or her disease. CT and MRI scanners allow doctors to 'slice' a living patient's brain without opening his or her skull. In this way, CT and MRI scanners have allowed doctors to visualise the cause of a patient's brain abnormalities since 1980. CT scans visualise the details of the body by means of X-ray images of thin slices of the body. MRI scanners use a strong magnet to achieve the same effect. CT scans hold the advantage of requiring less time inside the machine. It only takes several seconds or minutes to perform a CT scan, whereas MRI scans often take 20 min to complete. Another advantage of CT scans is that they allow for the easy identification of bones and bone fractures.

An example of a CT scanner, as well as an example of an MRI scanner, can be found on the next page.

© Springer Nature Singapore Pte Ltd. 2017
J. Hendrikse, *This is Our Brain*,
DOI 10.1007/978-981-10-4148-8_1

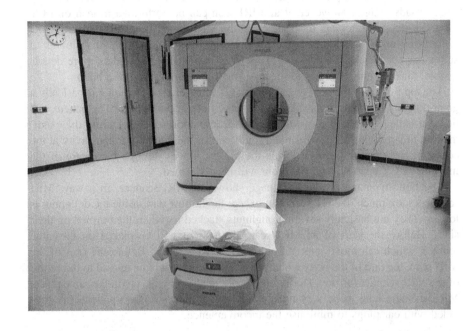

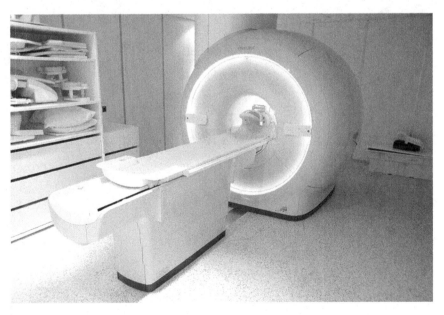

One of the disadvantages of CT is the use of X-ray technology, which poses certain small risks. In practice, the advantages of CT scans far outweigh the very slight disadvantages. However, if an MRI scan can be performed instead of a CT scan, doctors will often opt for performing the MRI scan. The disadvantage of MRI is that the tunnel is slightly longer, which may cause claustrophobic patients to suffer anxiety attacks. MRI is useful for brain scans since it renders more details visible, thus enabling doctors to identify small abnormalities in the brain.

The top image on the next page shows a CT scanner. A CT scanner is basically a giant doughnut turned on its side, with a large hole in the middle. The patient will lie on a table which will move during the examination (circle). During the examination, this moving table will slide a small part of the patient into the cavity at the centre of the scanner (arrow). While the patient moves through the centre of the machine, a CT scan of, say, the brain will be performed.

The bottom image on the next page shows an MRI scanner. In a way, MRI scanners look much like CT scanners, except the hollow part inside the doughnut is longer. They are basically several doughnuts stacked together, then turned on their sides. Before the start of an MRI scan, the patient will be placed on the table (circle), which can move into and out of the MRI scanner. Before the start of the MRI scan, the part of the body which must be visualised will be moved to the centre of the MRI scanner (arrow). During the scan, the patient will stay in the same place. MRI scanners tend to be very noisy while performing scans, so patients are provided with ear plugs to minimise the inconvenience.

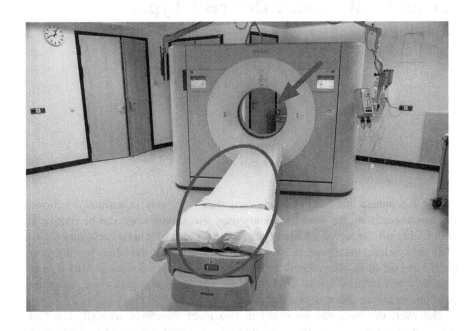

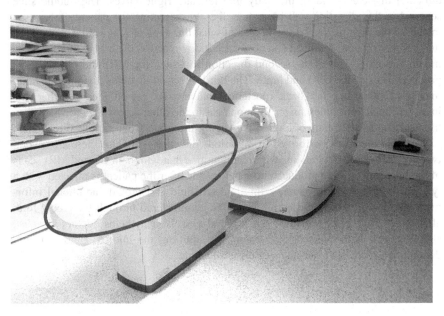

Chapter 2
CT and MRI Scans: Different Types of Images

Both CT scanners and MRI scanners can 'slice' the body in various directions. Horizontal scans are the most commonly used. Horizontal scans can be compared with slicing off the top of a boiled egg. To get an idea of what a horizontal CT or MRI slice is like, trace your index finger from one ear to the other, passing your cheeks and the tip of your nose along the way. This equals a horizontal MRI slice of your head and brain. When an MRI or CT scan of the head is performed, the entire head and brain are sliced in this direction.

However, the body can be sliced in two other directions, as well. The first direction involves dividing the body into left and right halves. The middle slice scanned in this direction will be a slice directly dissecting the patient's nose (see image on the next page). To get an idea of what this type of scan is like, place your index finger on your forehead and now trace it down across the bridge of your nose, mouth and chin. This is the second direction in which the MRI and CT scanners produce slices of the head and brain. The third direction involves dividing the body into front (anterior) and back (posterior) halves. To get an idea of what this involves, trace your index finger from one ear to the other across the crown of your head. CT and MRI scans often involve slicing the brain from various directions.

Not only are scans performed from various directions, but also there are several types of scans, as well. Patients will mainly notice that with both CT and MRI scans of the brain, images are produced both before and after the patient is injected with a type of fluid. In certain patients, this injected fluid can provide additional information. Generally, the fluid is intravenously administered through the elbow fold, after which it spreads over the body's blood vessels. This helps doctors visualise abnormalities in the vessels in question. In addition, the injected fluid can help doctors determine whether any vessels are leaking.

© Springer Nature Singapore Pte Ltd. 2017
J. Hendrikse, *This is Our Brain*,
DOI 10.1007/978-981-10-4148-8_2

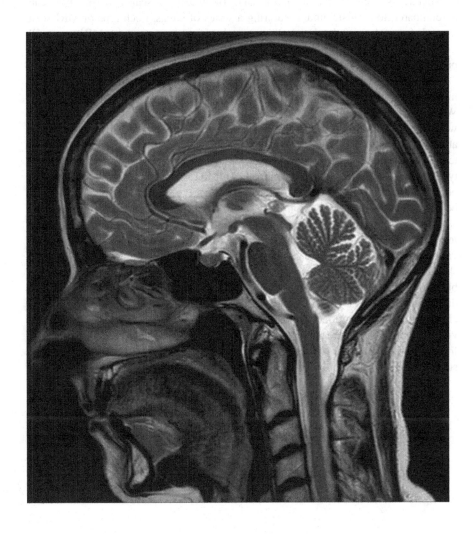

Normally, cerebral blood vessels are well insulated, so they do not leak. Some diseases of the brain may cause the vessels to springs leaks. If this is the case, the injected fluid leaks will be able to be seen in a CT or MRI scan of the brain. Scans made before and after the injection of the fluid will be compared to identify small leaks.

MRI scans often involve the performance of quite a few scans. An MRI scan can be compared to a music album featuring a series of songs. Each type of MRI scan takes 1–5 min to complete, and all songs taken together, a regular MRI scan will take about 15–30 min to complete. The various individual MRI scans will each provide unique information about the brain.

The image on the previous page shows an MRI image produced using a technique which is sensitive to water. Such MRI images are also known as 'scans'. When the brain is irritated, the amount of water may increase in places where one would not normally expect there to be much water. This type of MRI scan is good at visualising this. Pictured on the next page is an MRI scan produced using a technology which is sensitive to blood vessels. Over the years, dozens of clever types of MRI scans have been developed. An MRI scan of the brain often involves between 4 and 7 different types of MRI scan. What types of MRI scan a patient will be subjected to is determined on the basis of the patient's symptoms.

With CT scans, it is easier to try and remember what white, grey and black stand for. The various shades of grey are determined by the amount of X-ray radiation the various body parts block. Bones block a great deal of radiation. As a result, they show up white on CT scans. Air does not block any radiation; therefore, it shows up black on CT scans. Muscles and brain appear light grey, while fat appears a very dark grey. Water is a shade of grey somewhere between the shades of grey associated with muscles and fat. Although the order of greyness is fixed, it is sometimes vital that shades of white be rendered less white, and shades of black less black— for instance, to render small skull fractures more visible. In order to better visualise such fractures, the bone of the skull will be rendered less white.

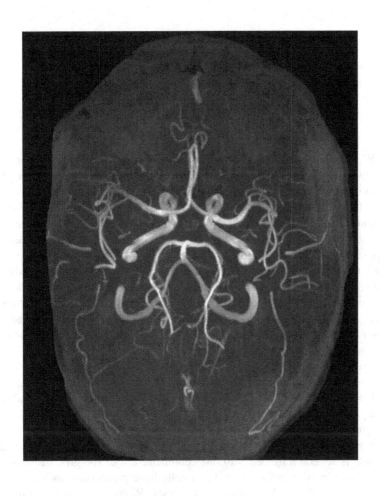

Chapter 3
The Anatomy of the Human Brain

The MRI scan on the next page shows an image in which the brain has been sliced right between the left and the right hemispheres. In order to get an idea of the direction in which this picture was taken, trace your finger from the middle of your forehead (between the eyes) down the tip of your nose and the middle of your lips to the middle of your chin. A light grey semi-arch can be seen at the centre of the brain. This semi-arch serves as a wide connection between the two hemispheres. Just above the arch, exactly in the middle of the brain, is a firm membrane which separates the two hemispheres. On the outer rim of the scan, a thin white edge can be spotted. This is the fat tissue situated just under the skin. The thin black edges which can be seen under the fat tissue are the outer and inner layers of the bones of the skull. Travelling further towards the centre of the brain, we will come across a thin layer of cerebrospinal fluid, which shows up dark grey. The brain can be found under this layer of cerebrospinal fluid.

It is easy to distinguish the two components of the brain on an MRI scan: the grey matter and the white matter. The names 'white' and 'grey' can be confusing, since, depending on the type of MRI scan performed, the levels of grey and white may vary, to the point where grey may become white and vice versa. In other words, grey and white matter do not always live up to their names.

Grey matter is positioned on the outside of the brain. The layer of grey matter on the outside is several millimetres to one centimetre thick. The grey matter houses the centre of the brain cells. These brain cells have projections, as well as connections to each other and to the rest of the body. White matter contains more fat compared to grey matter, as the nerves are protected by a fatty cover. Other areas of grey matter can also be found at the centre of the brain (called 'deep grey matter'). Many of these grey matter areas are connected to each other by means of axons.

© Springer Nature Singapore Pte Ltd. 2017
J. Hendrikse, *This is Our Brain*,
DOI 10.1007/978-981-10-4148-8_3

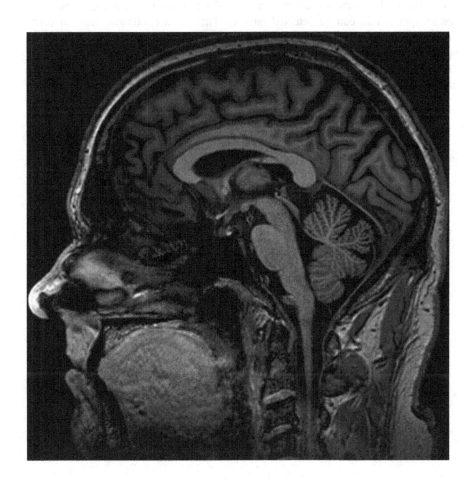

Thanks to a great number of interconnections, components of the brain are capable of taking over duties normally carried out by other components of the brain in the event of disease and/or brain damage.

The MRI scan on the next page shows the cerebellum at the bottom of the cerebrum. The brainstem can also be identified (circle). The midbrain is situated between the brainstem and the cerebrum. The midbrain and the brainstem form an important connection between the cerebrum and the cerebellum, and also the connection with the rest of the body, through the projections of the brain in the spinal canal. Between the cerebellum and the brainstem, a dark grey pool of cerebrospinal fluid can be detected (arrow). There is a continuous flow of cerebrospinal fluid from such pools to the cerebrospinal fluid around the brain and in the spinal canal. In order to prove that a patient is suffering from a particular disease, doctors sometimes need to suck up a small amount of cerebrospinal fluid through a needle. Generally, this fluid is sucked up from the spinal canal.

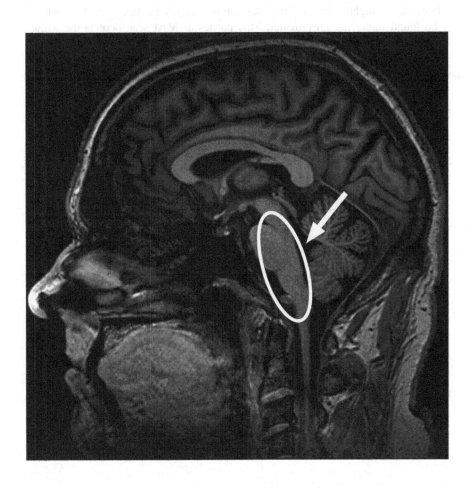

Generally speaking, doctors performing a brain scan will perform scans along a horizontal plane. The horizontal MRI scan on the next page shows the outer rim of the brain being slightly darker. This dark rim is the grey matter. The white matter can be found under the grey matter. This particular MRI scan depicts the white matter as being a lighter shade of grey. Deeper areas of the grey matter (circle) can be found towards the centre of the brain. These areas look dark grey, as well, as does the outer rim of the brain.

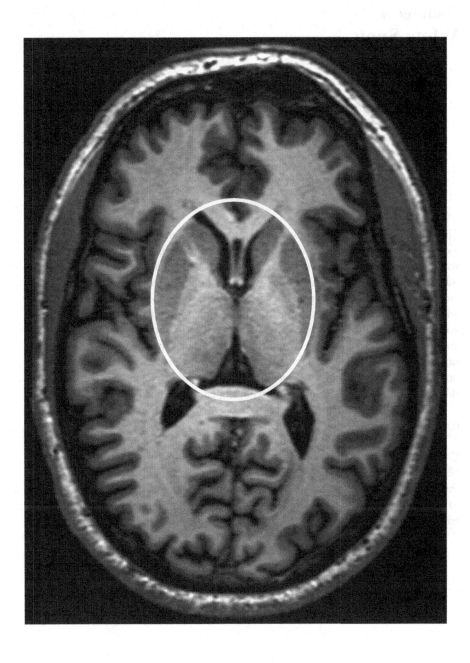

Chapter 4
White Spots

The image on the next page shows a horizontal MRI scan of the brain. This image is a cross section of a single slice, just a few millimetres wide. To give you an idea of what this is like, imagine the removal of the tip of the boiled egg you are eating on a Sunday morning. Suppose that you slide your knife horizontally through the egg and slice off the top of the egg. This will allow you to look inside the egg. The outer white layer of the egg can be compared to the brain's grey matter, while the yellow yolk inside the hard-boiled egg can be compared to the brain's white matter. MRI and CT scans allow one to peer into the brain in much the same way. Needless to say, no blood is spilled in doing so. Around 1980 is when we were first able to produce such cross sections of a patients' brain. Initially, CT scanners were used for this purpose; later, MRI scanners were developed, as well. CT scans generally take a few minutes to perform, whereas MRI generally requires 20–30 minutes to complete.

White spots can be seen in the cross section of the brain on the next page. Completely healthy young brains do not present with such white spots. They first become visible in many people from around age 40. Initially, they are small, but they will grow bigger with age. The white spots constitute slight damage (lesions) to the brain. They can be compared to wrinkles developing in a person's skin, or hair turning grey.

© Springer Nature Singapore Pte Ltd. 2017
J. Hendrikse, *This is Our Brain*,
DOI 10.1007/978-981-10-4148-8_4

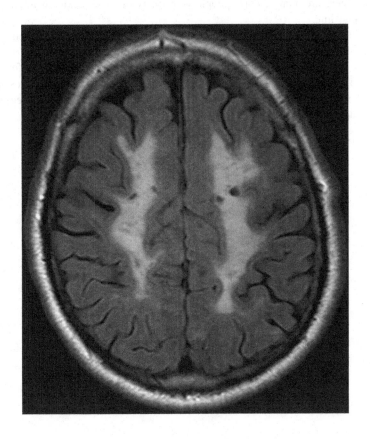

With age, more such small lesions will appear. Once we reach old age, nearly all of us will present with lesions, to varying degrees. Sometimes the abnormalities will grow so big that large, overlapping areas of white spots will develop (circles). If many components of the brain are thus affected, our memory may suffer. However, there are many elderly people with many of these lesions who appear to function just fine.

As with wrinkles, ageing is the main contributor to such damaged spots. However, there are other things which may contribute to the formation of white spots. For instance, high blood pressure will result in more white spots for some people. So far, no treatment has been developed for ageing brain. Wrinkles in the face can be treated with a facelift, but there is no 'brain lift' that will help people get rid of white spots.

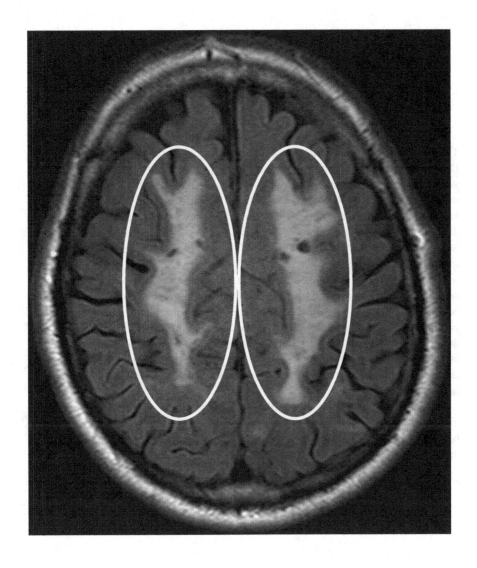

Chapter 5
Shrinking Brain

The image on the next page presents an example of brain shrinking with age. Doctors call this shrinking of the brain 'atrophy'. The shrinking process starts at a young age. Precise measurements will show a slight decrease in the size of the brain in persons aged 20 and older. Even though we still have to gain a great deal of knowledge and experience after the age of 20, we are expected to process and store this knowledge and experience in increasingly smaller brain. This generally works just fine; in many people, the shrinking of the brain is not noticeable until they are quite old.

On the MRI scan presented on the next page, the shrinking of the brain can be deduced from the increase in the amount of cerebrospinal fluid around the brain and between the brain folds (arrows). The cerebrospinal fluid shows up white. When people are young, their brains are often so large that there is limited space between the brain and the skull. As a result, the amount of cerebrospinal fluid between the brain and the skull will be limited in young people. Once people start ageing, the shrinking of their brain will increase the amount of room available between their brain and their skull.

If an ageing person's brain shrinks and simultaneously gets slightly damaged, they may suffer symptoms such as memory loss. The healthy components of the brain which still remain intact will not be able to completely compensate for such damage.

© Springer Nature Singapore Pte Ltd. 2017
J. Hendrikse, *This is Our Brain*,
DOI 10.1007/978-981-10-4148-8_5

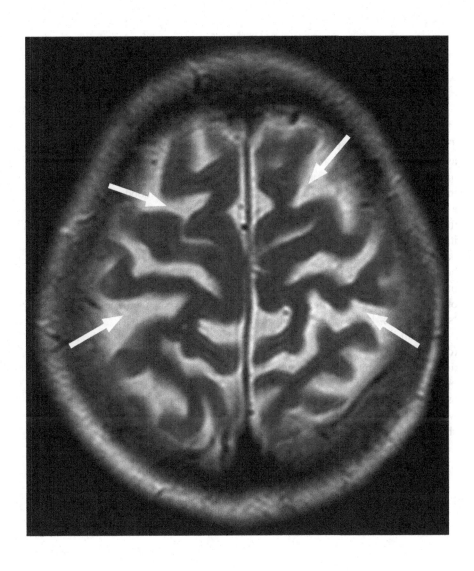

Chapter 6
Silent Infarcts

The image on the next page provides an example of a small area of dead brain tissue as shown on an MRI scan. The death of an area of the brain is called a cerebral infarction. Ageing people are increasingly prone to both severe and less severe infarctions. Generally, infarctions are caused by a blocked blood vessel. Since such a vessel cannot supply oxygen and nutrients to the brain, a part of the brain will die, thus causing an infarction. The more severe the infarction, and the more important its location, the more severe the patient's symptoms. If the infarction occurs in a part of the brain which has an important function, the patient is very likely to suffer health problems. Other parts of the brain can suffer an infarction without producing many symptoms. When doctors perform an MRI scan of a patient's brain, they will often chance upon evidence of previous infarctions. Such infarctions are called 'silent infarcts' because they result in few health issues, if any.

A small infarction, coloured white, can be seen in the image on the next page (circle). The brain tissue affected by the infarction has disappeared. Instead of brain tissue, a small white scar can now been seen. Cerebral infarctions can be identified on scans in two ways. For one thing, the brain tissue can be replaced by scar tissue, as in the image on the next page. This scar will be comparable to an old skin lesion which has left a visible scar. For another thing, a small cavity may develop at the site of the infarction. This cavity may or may not be filled with cerebrospinal fluid.

© Springer Nature Singapore Pte Ltd. 2017
J. Hendrikse, *This is Our Brain*,
DOI 10.1007/978-981-10-4148-8_6

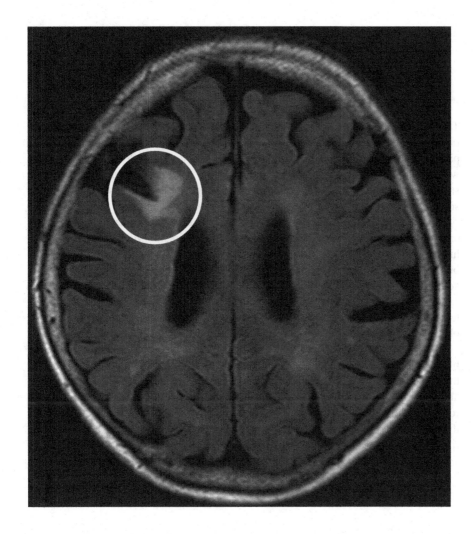

Chapter 7
Previous Microbleeds

The MRI image on the next page provides some examples of small microbleeds (areas with leftover blood) (pictured in black). Such microbleeds show that the patient has previously suffered from a cerebral haemorrhage. Some ageing people will experience the development of such bleeding spots in the brain. Doctors call such small haemorrhages or leftover blood as 'microbleeds' because they are very small compared with larger haemorrhages.

While large haemorrhages can be detected on CT scans, small haemorrhages cannot. The visualisation of small haemorrhages requires an MRI scanner which is highly sensitive to microbleeds. Such microbleeds may accidentally occur as the brain grows older. In addition, some patients may suffer large haemorrhages in addition to these microbleeds. Often, patients will undergo MRI to determine the cause of a large haemorrhage. Their MRI scans will show not just the large haemorrhage, but several older microbleeds. Microbleeds are also more commonly observed in patients with high blood pressure.

Generally, microbleeds can be identified on MRI scans because they show up black. In addition, these small areas with leftover blood can be recognised because they tend to be round. The round shape is caused by the actual blood in the middle and the effect of the old blood products on the MRI signal in its immediate surroundings. In other words, the black area identifiable on an MRI scan is larger than the actual bleed. Comparing it to a rock which falls into the water and produces ripples—the ripples are larger than the spot where the rock actually landed.

© Springer Nature Singapore Pte Ltd. 2017
J. Hendrikse, *This is Our Brain*,
DOI 10.1007/978-981-10-4148-8_7

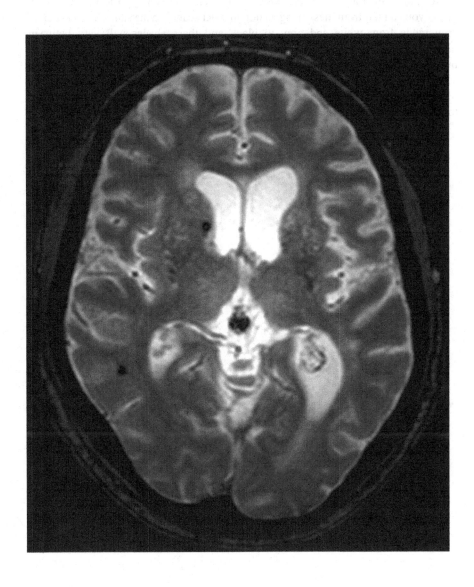

The MRI enlargement of a microbleed enables us to identify very minor bleeds (arrows). As the images on the next page show, in many cases there will be several types of abnormalities. In this case, the elderly patient's brain has shrunk and the brain is surrounded by thin white layer of cerebrospinal fluid. In addition, there are white spots which are consistent with small, ageing-related brain lesions (circles).

As you can tell from these images, not all MRI scans are the same. The shades of grey range from white and grey to black and these shades will vary between different type of MRI scans. As you will see, the cerebrospinal fluid surrounding the brain can be anywhere from dark grey to black to white, based on the type of MRI scan. The shades-of-grey settings can be adjusted to meet the requirements of the MRI scan, thus allowing the performance of MRI scans which are highly sensitive to, say, the microbleeds described above.

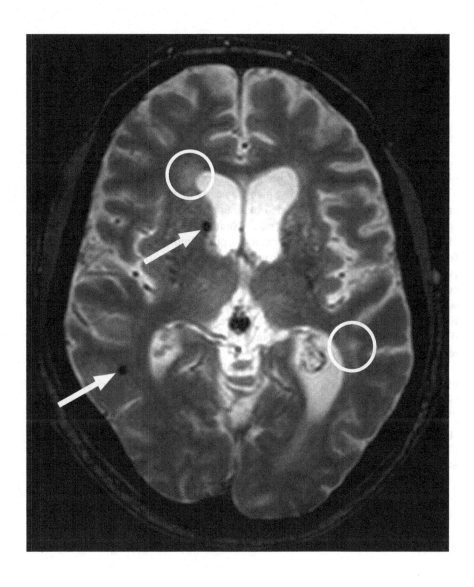

Chapter 8
More Room Around the Blood Vessels

The image on the next page shows white dots caused by an increased amount of cerebrospinal fluid surrounding the blood vessels in the deeper brain regions. These blood vessels go to the centre of the brain in order to supply oxygen and nutrients. They are very small blood vessels which are nearly impossible to spot in regular CT and MRI scans. They are surrounded by brain tissue. As we grow older, our brain will shrink a bit. As a result, more cerebrospinal fluid will flow between the brain close to the skull. In addition, more cerebrospinal fluid will flow between the brain folds. The brain do not just shrink from the outside; they shrink in all brain regions. As a result, a liquid rim will be visible around the blood vessels travelling to the centre of the brain. It is like putting on a coat. If you put on an oversized coat, there will be a lot of room between the coat and your body. In the brain, this excess room fills up with cerebrospinal fluid. On MRI scans, such fluid clearly shows up as a different shade of grey compared to brain tissue. As a result, the fluid can be readily identified on MRI scans.

On the scan presented on the next page, the cerebrospinal fluid surrounding the blood vessels can be seen in the form of several tiny white dots (circle). This is because, on this particular type of MRI scan, cerebrospinal fluid shows up white. The same white fluid can be seen on the outer rim of the brain, between the outer rim of the brain and the skull (arrows). The fact that more room is emerging around the blood vessels is often accidentally found when elderly patients undergo an MRI scan. As with shrinking of the brain, the degree to which the 'coat' around the blood vessels widens varies from person to person.

© Springer Nature Singapore Pte Ltd. 2017
J. Hendrikse, *This is Our Brain*,
DOI 10.1007/978-981-10-4148-8_8

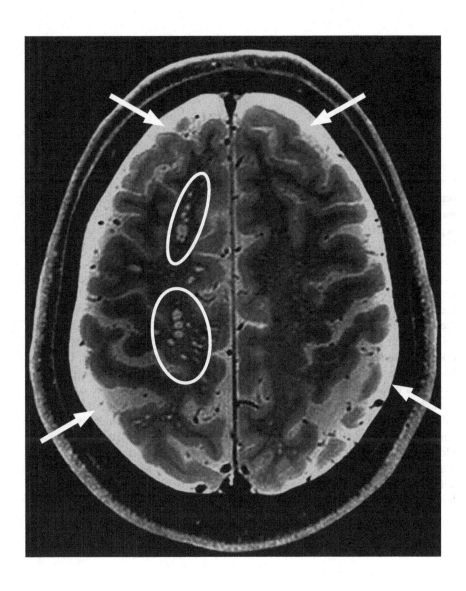

Chapter 9
Cerebral Infarction

The image on the next page shows an example of a major cerebral infarction. Cerebral infarctions occur when smaller or larger blood vessels are blocked. The brain requires the continuous supply of blood in order to receive oxygen and nutrients. When a blood vessel is blocked, the area of the brain which is normally supplied by this blood vessel will experience problems. The blood vessel shown here is a vessel supplying blood and oxygen to the brain, a so-called artery. Blood vessels which carry blood from the brain to the heart are called veins. If a blood vessel supplying blood to the brain (i.e. an artery) is blocked for a longer period of time, the brain tissue behind it will die, and this process is called infarction. If the blood vessel which is blocked is smaller, only a smaller part of the brain will die. If the blood vessel which is blocked is larger, a larger part of the brain will die. Large infarctions in key components of the brain will result in severe symptoms, such as difficulty moving the arms and legs, a drooping face and impaired speech. Damaged brain cells can attract and retain water, which may cause the brain to swell. In the event of a major infarction, the brain may swell to the extent that there is no longer enough room for them inside the skull. Some patients suffering from a major infarction will die. Patients who survive a major infarction will generally require a lengthy stint in rehab. There are types of medication which can eliminate the blockage of the blood vessel. In addition, the blockage can be sucked away through a tube inserted into the blood vessel. Both the medication and the suction must be administered quickly. The treatments are most effective if administered within a few hours. As English-speaking doctors have it, 'Time is brain.'

© Springer Nature Singapore Pte Ltd. 2017
J. Hendrikse, *This is Our Brain*,
DOI 10.1007/978-981-10-4148-8_9

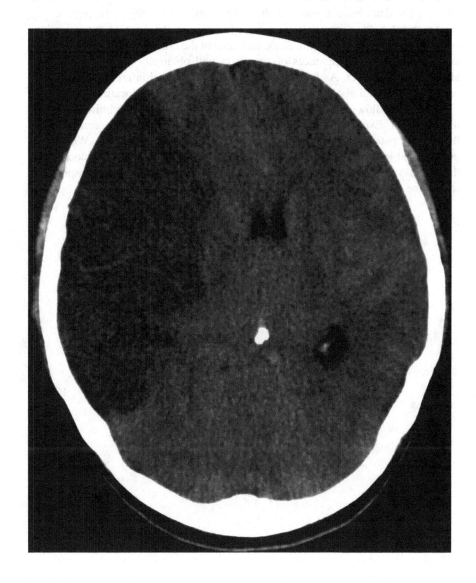

The image on the next page shows a major infarction (circle). What such infarctions look like on MRI and CT scans depends on the moment at which the scan is performed. During the first few days after an infarction, brain tissue will be swollen due to brain cell damage. In the months and years following the infarction, some of the dead brain tissue will disappear completely, only to be replaced by cerebrospinal fluid. There will often be scar tissue at the edge of the infarction, resembling a scar on one's skin. Since each area of the brain has its own unique task and function, the consequences of an infarct will obviously depend on the area affected by the infarct. As far as that is concerned, cerebral infarcts are like house prices. They are all about three things: location, location and location.

MRI scans allow doctors to detect swelling of the brain cells within minutes of the infarction. Eventually, the dead brain tissue will be replaced by cerebrospinal fluid. Some patients will suffer from a second infarction after the first. Whether or not a patient is at risk for a new infarction depends on the cause of the blockage of the blood vessel. Nearly, all patients require medications, including aspirin, to reduce the risk of new infarcts.

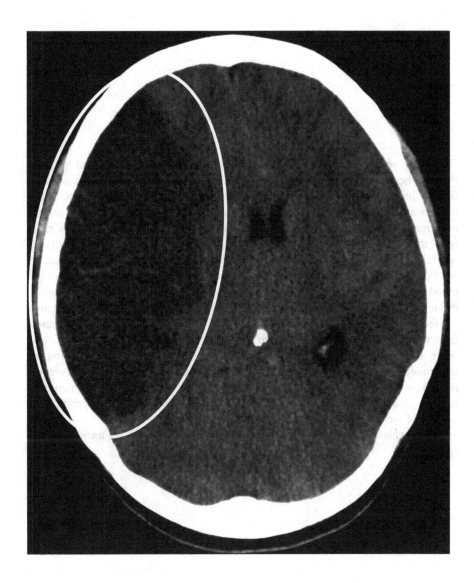

Chapter 10
Intracranial Haemorrhages

The image on the next page provides an example of a major intracranial haemorrhage. When an intracranial haemorrhage occurs, blood flows into the brain tissue from a blood vessel. Normally, blood remains inside the blood vessels while oxygen and nutrients are being exchanged with the surrounding brain tissue. An intracranial haemorrhage can be compared to a dyke breach, which allows river water to flood the hinterland. The brain tissue in the area suffering the bleed will get severely damaged; some tissue will die, while others will turn into scar tissue. In addition, blood products will remain in the brain tissue. The blood leaking from the blood vessel will turn into a big clot which will take up space. Since room is limited inside the skull, the formation of blood clots may mean that too little space remains for the brain inside the skull. This too limited room will result in an increased intracranial pressure, which, combined with severe damage to the brain itself, will prove fatal to some patients. Patients are at greater risk of dying following an intracranial haemorrhage than following a cerebral infarction. Like major cerebral infarcts, intracranial haemorrhages may result in permanent disabilities, which may be very severe. Here, too, symptoms vary widely between patients, depending on the site of the haemorrhage and the size of the bleed.

Haemorrhages which occur due to a dilation of a blood vessel form a special category.

© Springer Nature Singapore Pte Ltd. 2017
J. Hendrikse, *This is Our Brain*,
DOI 10.1007/978-981-10-4148-8_10

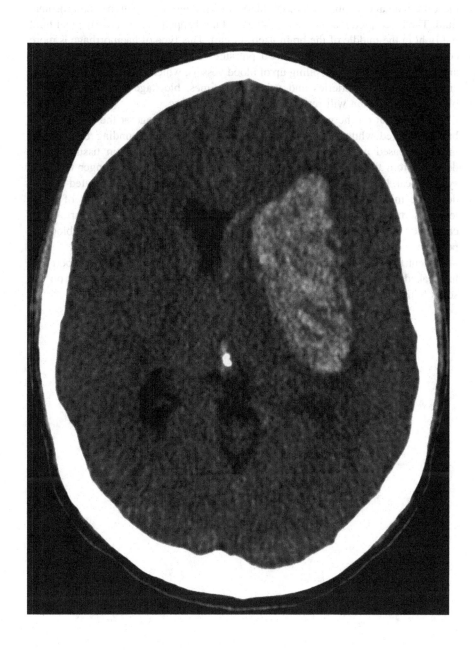

Blood vessels which carry blood from the heart to the brain are called arteries. Doctors call the dilation of such a blood vessel an aneurysm. If a patient suffers a bleed from an aneurysm, most of the blood will end up directly in the cerebrospinal fluid. The bleed pictured on the next page did not happen in the cerebrospinal fluid but right in the middle of the brain itself (circle). This type of haemorrhage is more common in patients with high blood pressure. Sometimes, it will be caused by another problem, such as a piling up of blood vessels, which develops due to a short circuit between the arteries and veins. Sometimes, blockage of a vein carrying blood from the brain will result in intracranial bleeding.

The CT scan on the next page shows a large haemorrhage at the centre of the brain, coloured white. This haemorrhage is putting the surrounding brain tissue under increased pressure. This pressure on the surrounding brain tissue can be deduced from the fact that the two cerebral hemispheres are no longer nice and symmetrical. Normally, the left and right cerebral hemispheres are separated exactly down the middle of the skull. However, due to the significant amount of blood leaked, some of this patient's brain is intruding on the other hemisphere (arrow). In cases such as this, some patients will undergo surgery to remove as much blood as possible.

Recent haemorrhages can be easily spotted on CT scans since fresh blood that has leaked from blood vessels into the brain tissue shows up bright white on CT scans (circle).

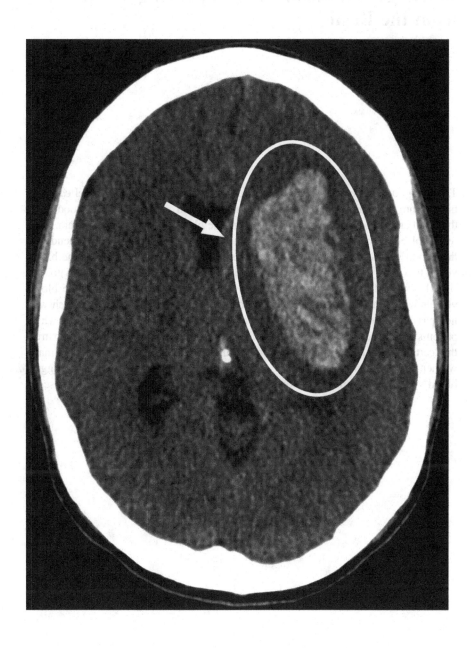

Chapter 11
Blockage of a Blood Vessel Carrying Blood from the Brain

The image on the next page shows an example of a clogged blood vessel carrying blood from the brain to the heart. Such clots in blood vessels carrying blood from the brain are dangerous because they prevent the blood from leaving the brain. As a result, bleeding may occur in the brain, in the part of the brain which depends on this blood vessel to remove blood from the brain. To illustrate, compare the blood vessel to a river which is suddenly dammed. If the water has nowhere else to go, the river banks will flood. When something like this happens in the brain, the blood vessels will start to bleed. Generally, patients will first suffer headaches which will not pass. Of course, heachaches have many causes, but every once in a while patients with headaches are actually suffering from a clogged blood vessel. It is true that certain patients are more susceptible to this than others. For instance, pregnant women and women who have just given birth are at increased risk for a clogged blood vessel. Sometimes, the blockage may be hard to spot on a CT or MRI scan.

© Springer Nature Singapore Pte Ltd. 2017
J. Hendrikse, *This is Our Brain*,
DOI 10.1007/978-981-10-4148-8_11

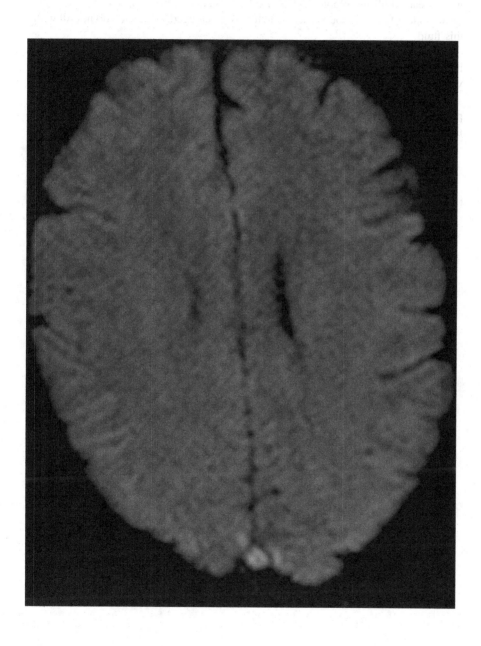

In order to render the clot more clearly visible, doctors will inject a special fluid into a blood vessel in the patient's elbow through an intravenous line. This fluid will spread to all the blood vessels inside the patient's body, which will help the doctors identify any blood vessels which may be clogged, since they do not fill with this fluid.

The light grey circle (arrow) on the MRI scan on the next page represents a clot in a blood vessel carrying blood from the brain to the heart. A healthy blood vessel would have shown up black in this type of MRI scan.

Patients with clogged blood vessels carrying blood from the brain will often be prescribed medication, which prevents blood coagulation (clotting) for a few months. This type of medication is called an anticoagulant.

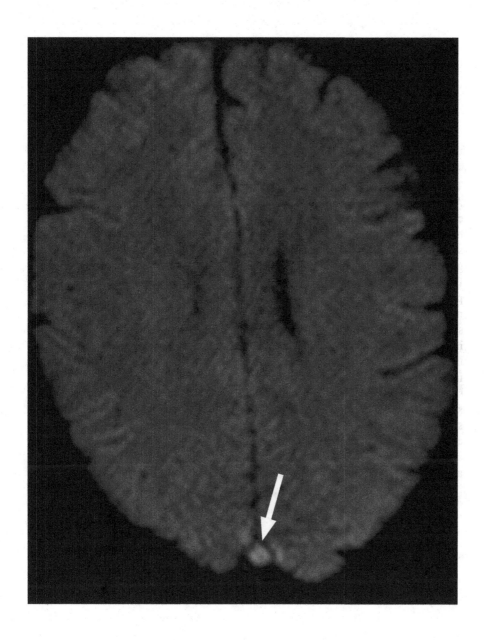

Chapter 12
Blocked Paranasal Sinuses

The image on the next page shows an example of blocked paranasal sinuses. Admittedly, the sinuses are not part of the brain itself, but they are situated very close to the brain. Therefore, blocked paranasal sinuses will sometimes result in headache. Partially blocked sinuses are often accidentally found on scans like MRI scans. Many patients do not have any symptoms. Blocked paranasal sinuses are generally caused by an inflammation of the mucous membranes of the sinuses. Doctors call this condition 'sinusitis', with '-it is' meaning 'inflammation'. When people have a cold, their sinuses are often partially clogged. In many cases the mucous membrane will only thicken along the walls of the sinuses. Sometimes the sinuses will fill up with fluid. When a patient undergoes X-rays or a scan, this fluid in the paranasal sinuses will show up as a horizontal surface. The air contained in the sinuses will be visible above the surface, and the fluid which built up due to the inflammation will be visible below the surface. Completely clogged paranasal sinuses may involve the entire air filled sinus being filled up with a combination of swollen mucous membranes and built-up fluid. People have several paranasal sinuses. If one sinus is clogged, the other sinuses will often be involved to some extent as well. Some people suffer chronically from blocked sinuses. These people can choose to undergo surgery to widen the space between the paranasal sinuses and the nasal cavity, thus allowing fluid to drain from the sinuses more easily.

© Springer Nature Singapore Pte Ltd. 2017
J. Hendrikse, *This is Our Brain*,
DOI 10.1007/978-981-10-4148-8_12

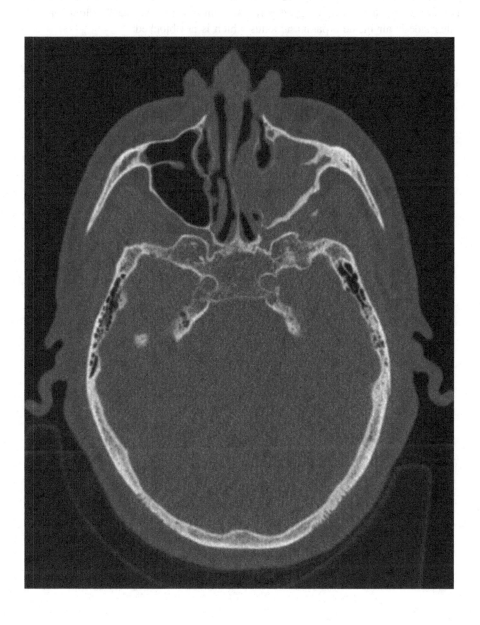

The CT scan on the next page shows a blocked paranasal sinus, whose fluid shows up grey (circle). The white lines are the bone surrounding the sinus. On the other side, the centre of the sinus is much blacker (arrow). On CT scans, air shows up black, water shows up dark grey, and bones show up white. Therefore, the grey colour on one side is the clogged paranasal sinus, while the black colour on the other side is air inside a paranasal sinus which is not blocked.

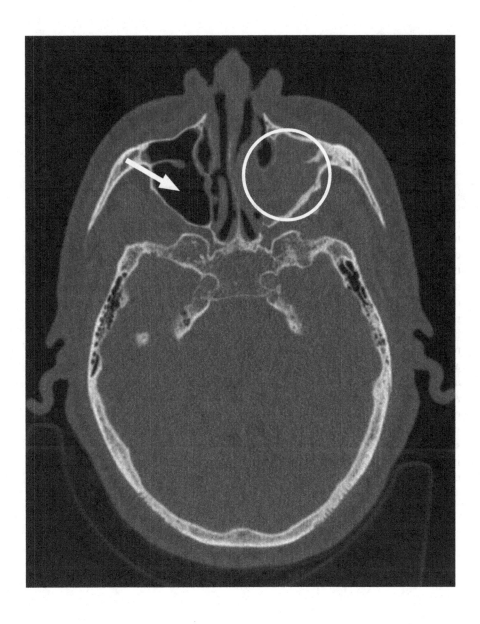

Chapter 13
Multiple Sclerosis

The image on the next page is an MRI scan of a patient with multiple sclerosis (MS). CT scans are not suited to detect the kind of abnormalities which can be observed in MS patients. In MS patients, the fatty insulating sheaths surrounding the neural pathways are damaged. These fatty insulating sheaths allow signals transmitted by the brain to travel uninterrupted. Doctors call MS a 'demyelinating disease' because it reduces the amount of myelin, the fatty substance surrounding the nerves. Despite a great deal of research, scientists have not yet been able to determine what causes MS. However, what is clear is that an inflammatory reaction occurs in the insulating sheaths surrounding the neural tracts. The inflammatory reaction is progressive over time. Typical periods of relative calm will be followed by periods with new inflammatory reactions. Over the course of several years, these recurring inflammations will cause the insulating sheaths to get increasingly damaged, which will result in the MS patient experiencing more and more symptoms and impairments. The optic nerve, too, is surrounded by such an insulating sheath. As a result, impaired vision is sometimes the first symptom of an MS patient. MS is often diagnosed because of recurring, increasingly severe symptoms, interspersed with periods of relative calm. The disease is not always easy to diagnose, and MRI scans are a major source of supplementary evidence. Due to the lesions on the insulating sheaths surrounding the nerves, MS often presents quite a distinct look on MRI scans.

© Springer Nature Singapore Pte Ltd. 2017
J. Hendrikse, *This is Our Brain*,
DOI 10.1007/978-981-10-4148-8_13

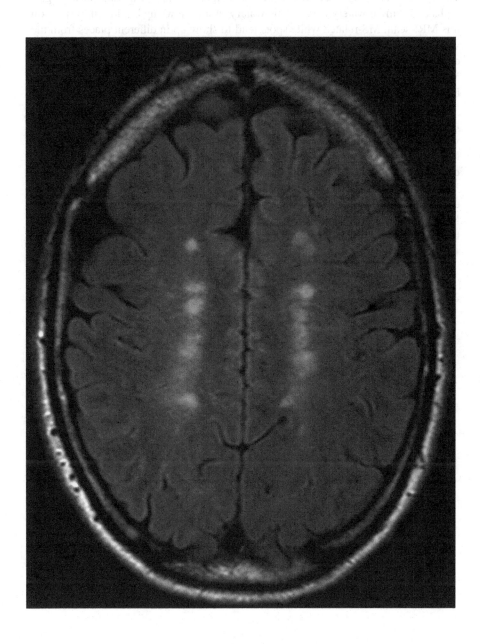

The MRI scan pictured on the next page shows areas with white spots (circles). These white spots develop when the fatty signal of the insulating sheaths gets replaced by more watery tissue. This watery tissue shows up bright white spots on this MRI scan. MS-related white spots tend to show up in different places from the white spots normally found in ageing brains. In addition, MS patients tend to be quite young, so one would not expect to see a large number of white spots. MS-related white spots tend to be located very close to the small pools of cerebrospinal fluid which can be found at the centre of the brain. These pools are called 'ventricles'. There are other places where MS-related white spots are often found. It is believed that these places are affected in MS patients because they are the places which are most easily accessed by the body's inflammatory cells. Generally, MS-related white spots are first observed in the areas surrounding the blood vessels carrying blood from the brain to the heart (veins).

Analysis of patients' symptoms and impairments and MRI scans can help doctors diagnose patients with MS, or alternatively rule out MS. Generally, several MRI scans will be required to determine whether new white spots appear in the brain, which constitutes additional proof of MS.

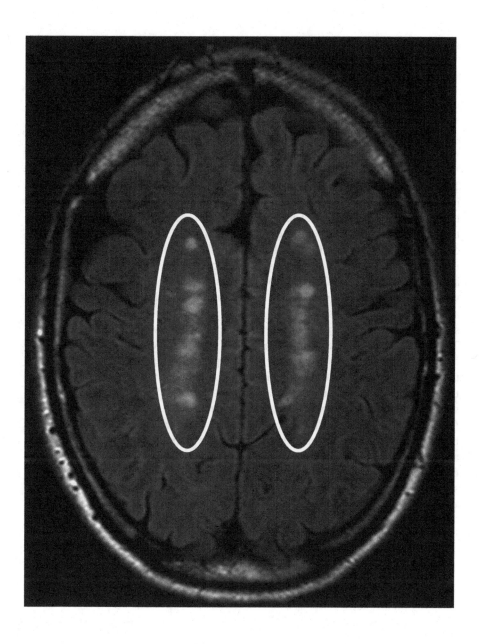

Chapter 14
Bulging Intervertebral Discs

The image on the next page shows an MRI scan of a patient with a bulging intervertebral disc. We have now left the brain and gone south a bit. Nerves exit the brain to enter the spinal canal inside the spine. Nerve roots exit the spinal canal through small openings at both sides of the spinal canal. The spine supports the rest of the body like the trunk of a tree. In addition, the vertebrae protect the nerve roots inside the spine. A great deal of pressure is exerted on the bottom of the spine since all the building blocks of the vertebrae are supported by it. As we age, or if our work involves hard physical labour, the bottom of our spine will experience wear and tear. The spine consists of 24 vertebrae stacked on top of each other. Little shock absorbers called intervertebral discs are located between all these vertebrae. The intervertebral discs constitute the flexible cement between the vertebrae. They allow the spine to move, for instance when a person bends over. Wear and tear of the spine is typically first observed in the discs between the vertebrae. Normally, these intervertebral discs contain a fair amount of water. When the spine has been subjected to wear and tear, this water disappears and the intervertebral discs grow flatter. When a disc grows flatter, a section of it may start bulging. If the resulting bulge touches a nerve, the patient will experience symptoms which will affect the area of the leg served by the compressed nerve. As a result, doctors can sometimes determine quite easily which nerve is being compressed by an intervertebral disc, simply on the basis of the patient's symptoms.

© Springer Nature Singapore Pte Ltd. 2017
J. Hendrikse, *This is Our Brain*,
DOI 10.1007/978-981-10-4148-8_14

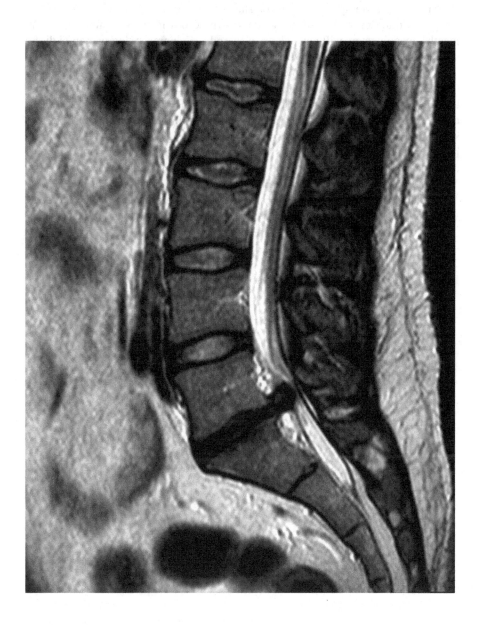

Bulging intervertebral discs are also known as 'hernias'. As we grow older, we will often develop hernias which do not cause any nerve pain. When a patient is experiencing severe symptoms, he or she may be operated on. However, since symptoms often grow less severe of their own accord in the next few weeks or months, many doctors will now take a 'wait-and-see' approach. Intervertebral discs shrink over time, thus causing the symptoms to become less severe.

The MRI scan on the next page depicts a bulging disc which shows up black (circle). MRI scans are best suited to visualising intervertebral discs. Above this bulging disc, square vertebrae can be detected. This MRI scan shows a body bisected at its exact core, presenting us with a slice several millimetres thick, seen from the side. For a better understanding of what you are looking at, imagine a body which has been cut into two halves with a sword without shedding any blood in one smooth movement, from the top of the head through the nose and further down the body. You are now looking at one slice from the middle of the body, seen from the side.

In addition to bulging discs, patients often suffer some damage in the adjacent vertebrae. Wear often affects several vertebrae and intervertebral discs simultaneously.

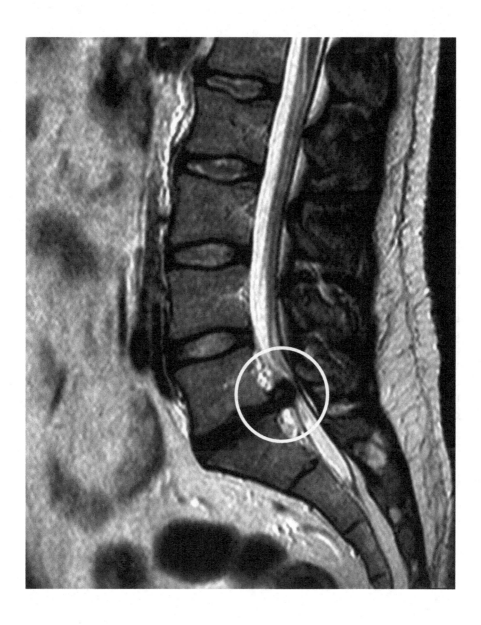

Chapter 15
Spinal Stenosis

The image on the next page shows a CT scan of the upper part of the spine. As we grow older, the upper half of the spine will experience great wear and tear. This wear can be seen in X-ray images or scans in the form of hook-shaped deformities, which are small projections at the edge of a bone. In this case, a vertebra. When we are young, our vertebrae start out looking like square building blocks. Yet in the course of time, they will develop small projections on their outer edges, e.g. where they meet the adjacent intervertebral discs. These projections are known as 'bone spurs'. They are hook-shaped deformities which often present alongside worn intervertebral discs. If the hooks grow large, they may cause the spinal canal to grow narrower, which is especially common in the top and bottom parts of the spine. The middle part of the spine seems much less prone to such narrowing.

The brain has a projection of its own inside the spinal canal. It is called the 'spinal cord'. Nerves branch off from this spinal cord. For most people, the spinal cord is located in the first twenty vertebrae of the spine. Below the 20th vertebrae, the spinal canal only contains nerves. Nerves branch out from the spinal canal on either side of the spine between the vertebrae.

© Springer Nature Singapore Pte Ltd. 2017
J. Hendrikse, *This is Our Brain*,
DOI 10.1007/978-981-10-4148-8_15

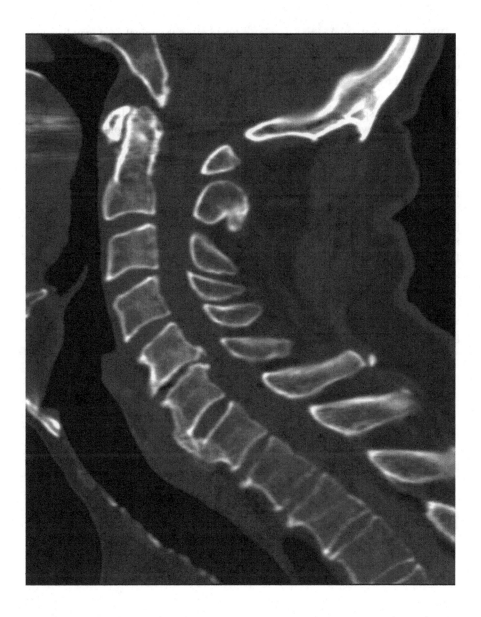

When the spinal canal grows narrower due to wear, the spinal cord may become constrained. Normally, the spinal cord is situated right in the middle of the spinal canal, but bone spurs may cause it to be pressed backwards or to the sides. Bone spurs caused by wear and tear may also cause the gaps from which the nerve roots exit to narrow, which means that, as with bulging discs, severe wear involving the growth of bone spurs may result in a nerve getting compressed. If the spine has sustained severe wear and tear, patients may suffer both bone spurs and bulging discs, which will have a cumulative effect. If the spinal canal suffers severe stenosis (narrowing) and if the spinal cord is badly compressed, patients may want to undergo surgery to widen the part of the spinal canal affected by stenosis. This often involves removing parts of several vertebrae at the rear of the spine.

On the CT scan on the next page, the vertebrae show up white. The sharp projections at the front and rear of the vertebrae are bone spurs (circle). These bone spurs were caused by wear and tear. This CT scan shows the vertebrae scanned from the side. The hooks at the back of the vertebrae have caused the dark grey area directly behind the vertebra to grow narrower. The dark grey area is the spinal canal (arrows). It is causing the projection of the brain in the spinal canal (i.e., the spinal cord) to be compressed. The CT scan also shows a bone spur at the front of the spine, between two vertebrae (dotted circle). Bone spurs at the front of the spine generally do not cause any symptoms.

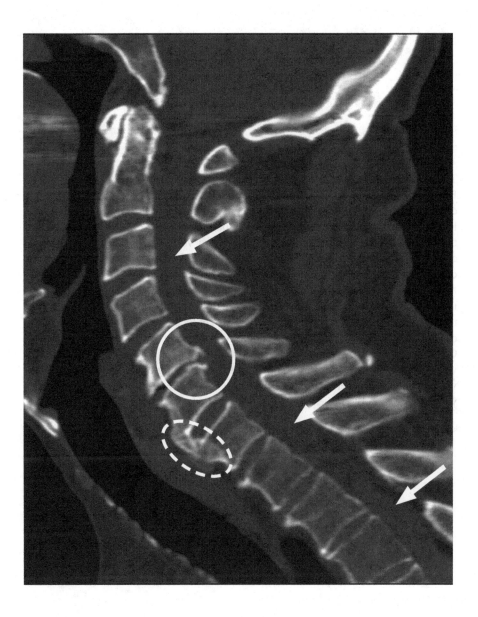

Chapter 16
Blood Vessels of the Neck

The MRI scan on the next page shows the arteries which carry blood to the brain. Blood vessels which carry blood to the brain are called arteries, while blood vessels carrying blood from the brain to the heart are called veins. The largest blood vessel in the human body is the aorta. The aorta has several branches at the top of the thorax which go to the neck region. These blood vessels each have several branches, e.g. the blood vessels which carry blood to the face, tongue, skin and muscles of the neck. The largest blood vessels carrying blood to the brain pass through the front of the neck, just off the centre of the neck. One can often feel one's pulse in these blood vessels. So just like one can check one's pulse at the wrist, one can do this also in the neck. Since these blood vessels are located close to the surface, they can be inspected using ultrasound technology; it does not always require an MRI or CT scan. As we grow older, nearly all of us will experience a thickening of the walls of the blood vessels. In some people, this thickening of the walls of the blood vessels will cause severe narrowing in certain spots. The disease of the vessel walls which causes the blood vessels to narrow is called atherosclerosis.

© Springer Nature Singapore Pte Ltd. 2017
J. Hendrikse, *This is Our Brain*,
DOI 10.1007/978-981-10-4148-8_16

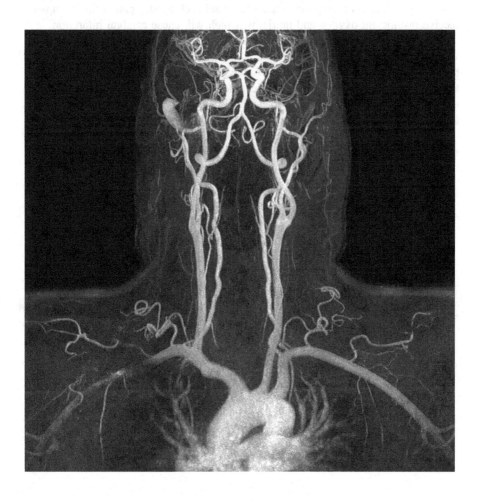

Sometimes, the vessel walls will be so diseased that they become completely clogged. It goes without saying that this may jeopardise the supply of blood to the brain. In addition to complete clogging, there is another problem, namely the fact that small pieces of diseased vessel wall may break off and enter the blood stream. These small pieces will travel through the blood vessels downstream towards the smaller blood vessels in the brain. If these small pieces get stuck inside small blood vessels, these vessels may get clogged. If a vessel is clogged, a part of the brain will receive insufficient oxygen and nutrients, which will cause cerebral infarction.

The MRI scan on the next page shows the various blood vessels of the neck, which show up white on the scan. At the bottom of the image, the blood vessels of the neck can be seen to branch off from the aorta (circle). As they move upwards, the blood vessels branch into multiple other vessels, which progressively grow thinner. At the top of the MRI scan, the blood vessels have entered the skull (arrows), where they will branch off into many smaller vessels.

It is vital that the blood vessels carrying blood to the brain do not have a sudden significant change in size. A sudden narrowing or clogging of a blood vessel in the neck region is one of the possible causes of a cerebral infarction.

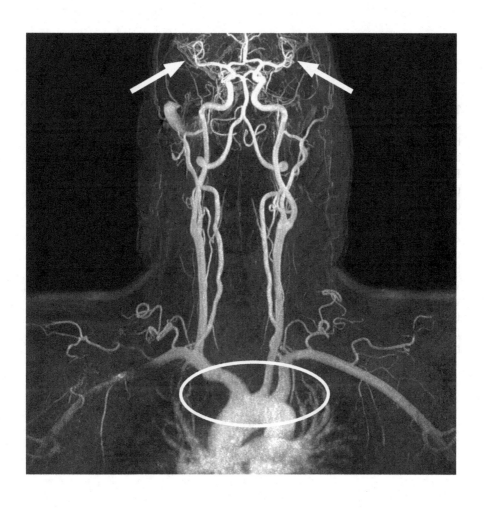

Chapter 17
The Intracranial Blood Vessels

The MRI scan on the next page shows the intracranial blood vessels inside the skull. The blood vessels are the motorways along which blood is transported to the brain. There are six such motorways inside the skull. Both sides of the skull have an anterior, middle and posterior blood vessel, which makes six blood vessels in total. Blood is carried to these six intracranial blood vessels through the blood vessels in the neck region, which distribute the blood over these six blood vessels. This distribution occurs deep inside the skull, on something which looks like a round-about. Since this roundabout of intracranial blood vessels was discovered by Sir Arthur Willis, it is called the 'circle of Willis'. Not far from this circle, the anterior, middle and posterior blood vessels split into smaller blood vessels, which in turn branch into even smaller blood vessels, until the blood finally reaches the brain. The very smallest blood vessels pass on oxygen and nutrients to the brain, and remove waste. They then merge again to carry blood from the brain to the heart, growing larger as they travel away from the brain. Finally, the blood vessels carrying blood from the brain merge into two large blood vessels in the neck, which carry the blood back to the heart. Each of these blood vessels can get clogged.

© Springer Nature Singapore Pte Ltd. 2017
J. Hendrikse, *This is Our Brain*,
DOI 10.1007/978-981-10-4148-8_17

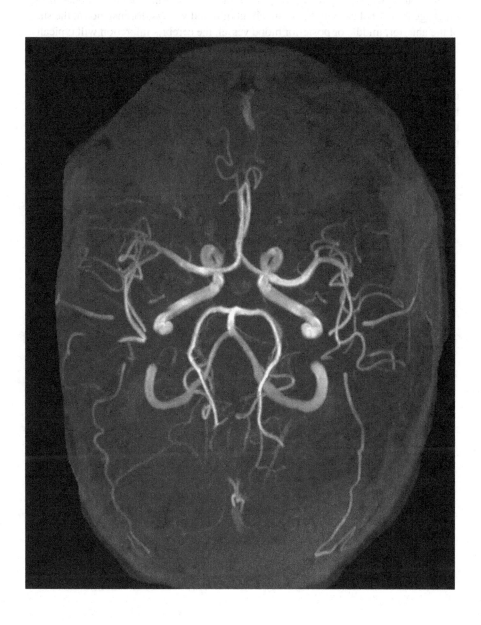

The most common type of blockage occurs in the intracranial blood vessels carrying blood to the brain. When such a blockage occurs, the brain downstream will not receive sufficient blood. Without oxygen and nutrients, the brain will quickly get into difficulty, which may result in cerebral infarction within minutes or hours. If the blockage is situated close to the roundabout of blood vessels, for instance at the start of the anterior, middle or posterior blood vessel, the cerebral infarction will typically be severe. Blockages situated farther downstream in very small blood vessels carrying blood to the brain will generally result in small infarctions.

The MRI scan on the next page shows the intracranial blood vessels, which show up as white stripes. The roundabout of blood vessels called the 'circle of Willis' can be seen at the centre of the scan (circle). The six main motorways which transport the blood to the brain (anterior, middle and posterior blood vessels on each side of the skull—arrows) branch off from this roundabout, which is situated deep inside the skull. These blood vessels branch into smaller blood vessels not long afterwards.

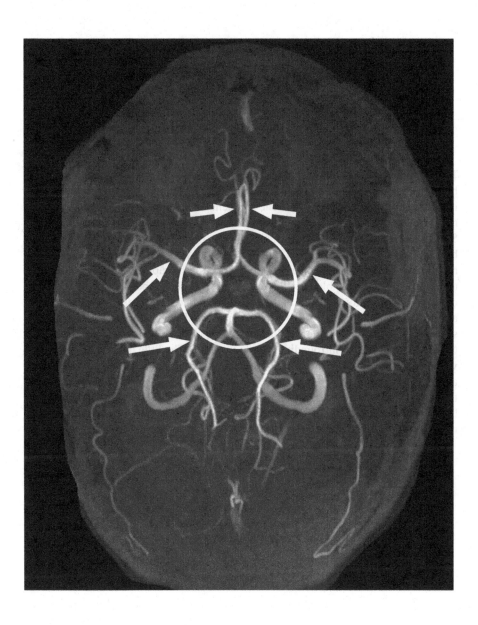

Chapter 18
Ruptured Aneurysm

The CT scan on the next page shows the intracranial blood vessels supplying the brain with blood. These blood vessels sometimes rupture. Such ruptures will immediately cause the worst headache a patient has ever had. When a blood vessel ruptures, a weak spot can generally be observed in it at the location of the rupture. Such weak spots often look like small balloons on the blood vessel. Just like real balloons, such small balloons on blood vessels may rupture. When a real balloon ruptures, a loud bang can be heard at the moment of the rupture. When a balloon on a blood vessel ruptures, blood will leave the vessel. The blood vessels, like the rest of the brain, are surrounded by cerebrospinal fluid. In other words, the blood vessels are small tubes swimming in the cerebrospinal fluid which surrounds them on all sides. So when a weak spot on a blood vessel ruptures, a lot of blood will flow into the cerebrospinal fluid surrounding the blood vessel. The little balloon on the blood vessel which is generally the cause of such a rupture is called an 'aneurysm'. Such aneurysms occur at several specific locations on the blood vessels, particularly in those parts of the vessels where they branch into smaller vessels. Many such branches, which are preferred locations of aneurysms, are located around the roundabout of blood vessels deep inside the skull.

© Springer Nature Singapore Pte Ltd. 2017
J. Hendrikse, *This is Our Brain*,
DOI 10.1007/978-981-10-4148-8_18

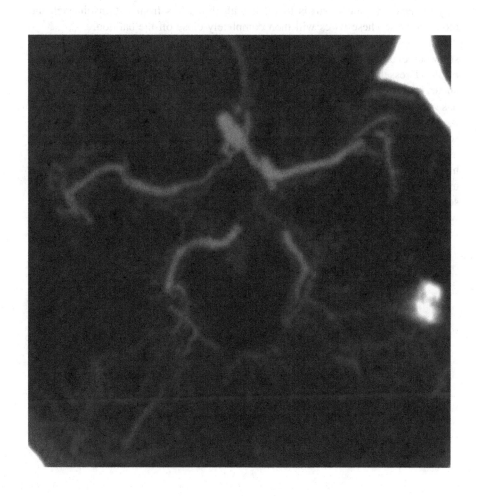

The reason why aneurysms occur at branching locations is blood vessel walls are less strong at bifurcations. As a result of this weakness, a small balloon-like bulge may develop. The bigger the balloon, the greater the risk of a rupture. Some patients will die after a ruptured aneurysm due to the consequences of severe bleeding. In other patients, the bleeding will stop. Even after the bleeding has stopped, patients are still at increased risk of more bleeding. In such cases, doctors may surgically insert a tiny 'clip' on the balloon to ensure that it does not rupture again. Another way to repair the aneurysm is to fill it with thin wires from within, through the blood vessels. These wires will then completely close off the balloon.

On the CT scan on the next page, the blood vessels show up white. The skull shows up even brighter white than the blood vessels. At the front of the roundabout of blood vessels, a small balloon can be seen (circle). Generally, such balloons, or aneurysms, are smaller than 1 cm. Around 3 percent (3%) of the human population has a weak spot in an intracranial blood vessel that has not ruptured. Sometimes these weak spots are accidentally detected when a patient undergoes an MRI or CT scan for another reason. In many cases, patients with dilated blood vessels will be asked to return for a scan one or several years later to check whether the dilations have stayed the same size or grown bigger. When such dilations are treated, the treatment itself brings a small risk of a cerebral infarction. As a result, it is sometimes more sensible not to treat a patient when a small aneurysm is present which has not ruptured.

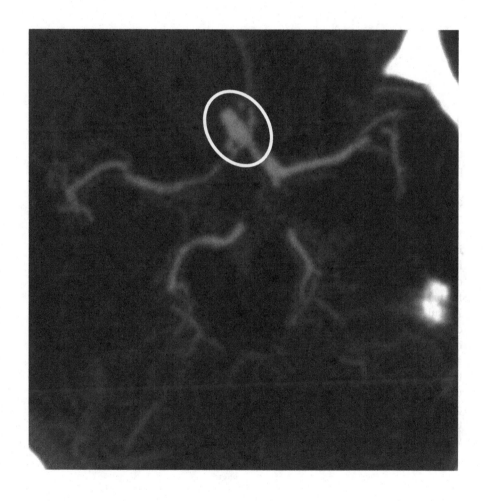

Chapter 19
Skull Fractures

The CT scan on the next page shows a skull fracture. The skull is an important part of the body because it protects the brain from external hazards. The skull prevents our brain from getting damaged every single time we bump our heads. Thanks to the thickness of the bones of the skull and the skull's round shape, our brain is well protected. However, sometimes the skull will receive such a blow that it will break. Just like the shell of an egg may crack, so the human skull may crack. Skull fractures are generally caused by accidents, particularly road accidents. Other accidents which may involve skull fractures include falls from great height, e.g. a fall from the stairs or from a ladder. Alternatively, accidents occurring while carrying out one's work or indulging in a hobby (for instance, a fall from a horse while not wearing a helmet) may be to blame. The location and severity of the skull fracture depend on the nature of the accident. Road accidents often involve multiple fractures, including skull fractures. Skull fractures may coincide with fractures in the maxillofacial bones. A skull fracture does not have to pose a danger. However, it does often involve a concussion. Skull fractures pose a hazard if the fracture causes a haemorrhage between the skull and the brain. Bleedings between the skull and the brain can rapidly grow so severe that the blood will start pressing on the brain. By opening a small trapdoor into the skull (a so-called bone flap) in good time, neurosurgeons can stop the bleeding and stop the compression of the brain by the bleeding. Such bone flaps also enable surgeons to remove the blood which is pressing on the brain. In addition to causing fractures, accidents may damage the brain itself due to the force with which the brain meets the skull.

© Springer Nature Singapore Pte Ltd. 2017
J. Hendrikse, *This is Our Brain*,
DOI 10.1007/978-981-10-4148-8_19

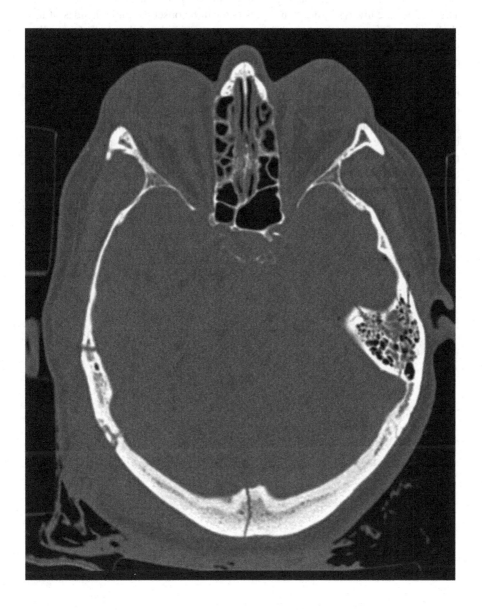

The CT scan on the next page shows several skull fractures (circles). The bones of the skull show up white on the CT scan, while the fracture shows up dark grey. The skull consists of several bones which are tightly held together. It is vital that the edges of the various bones making up the skull are not be mistaken for fractures. These edges are always in the same spots and can be observed on both sides of the skull, whereas fractures tend to show up on just one side of the skull. Generally, skull fractures are located at the exact same spot where the patient's skin is swollen due to the impact of the accident. This makes it easier to identify skull fractures.

Skull fractures may result in rupturing of blood vessels between the skull and the brain, which will cause blood to pool between the skull and the brain. Such pooling of blood between the skull and the brain increases the risk of dangerous compression and displacement of the brain, which may cause severe damage to the brain. Therefore, it is crucial that bleeding between the brain and the skull is recognised in time, so that the blood can be surgically removed by a neurosurgeon. Such an operation involves making a small bone flap in the bones of the skull through which blood can be removed and the bleeding stopped.

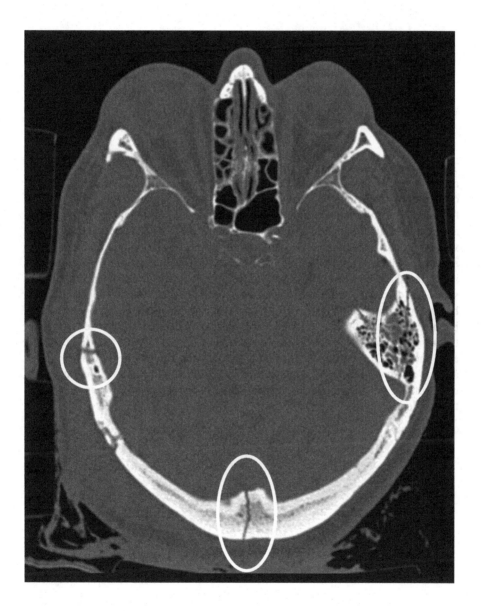

Chapter 20
Post-accident Haematoma

The CT scan on the next page shows a brain haematoma which occurred after an accident. A haematoma can be compared with a severe bruise or black eye, only it occurs inside the brain tissue. If the skull experiences great trauma, for instance due to a road accident, the brain may be severely damaged. What happens during road accidents or falls from great heights is that the skull will suddenly come to a halt, while the brain inside the skull is still moving. As a result, the brain will be flung against the inside of the skull. The impact will cause the brain to move up and down inside the skull for a brief moment, which will often damage the brain in multiple spots. The damaged spots tend to be located opposite to each other. The damage can be observed on a CT or MRI scan, where it will look like a severe bruise of the brain tissue. Often, the extent of the damage can only partially be detected just after the accident, but will be more visible when the CT scan or MRI scan is repeated after one day. The consequences of brain haematomas can still be seen many years after an accident as they will leave scar tissue. Also, blood breakdown products left over from a brain haematoma after an accident can be seen on MRI scans years after the accident.

© Springer Nature Singapore Pte Ltd. 2017
J. Hendrikse, *This is Our Brain*,
DOI 10.1007/978-981-10-4148-8_20

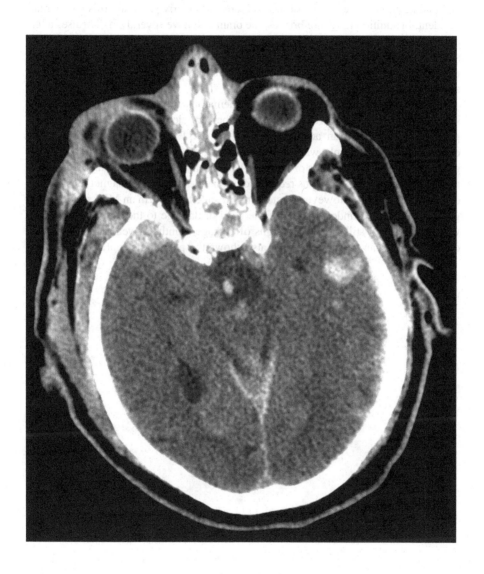

There are several places which are prone to bleeding after an accident. The parts of the brain adjacent to the bottom of the skull are the most susceptible to post-accident bleeding because the force exerted on the brain due to its movement inside the skull is the greatest here. The sections of the brain adjoining the anterior and middle parts of the bottom of the skull are particularly prone to bruising after an accident. In addition to severe bruises, the brain will have several small bruises after a severe accident. MRI scans tend to be better than CT scans at picking up such small bruises, since MRI scans are more sensitive to small amounts of blood. If there are many small bruises spread all over the brain, it is a sign that the brain is severely damaged.

The CT scan on the next page shows a haematoma adjacent to the bottom of the skull (circles). Since fractures and blood can be very clearly identified on CT scans, people who have had an accident tend to get a CT scan. Moreover, time is of essence in treating people who had a severe accident, and performing a CT scan generally takes a few seconds or minutes, whereas an MRI scan may take up to half an hour. In addition, MRI scans require additional care since the magnet may attract certain metal objects. However, doctors may decide to perform an MRI scan later to get a more detailed understanding of the level of damage to the brain, or to check whether the brain still shows any signs of damage months or years after the accident. The CT scan on the next page also shows the patient's eyes (arrows).

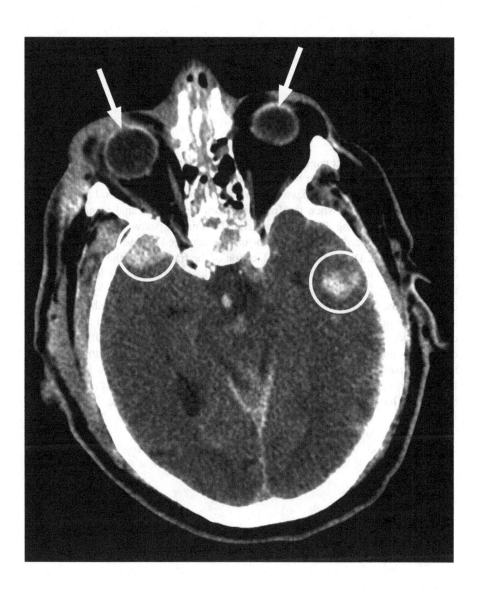

Chapter 21
Vertebral Fracture

The CT scan on the next page shows a fracture in a vertebra in the upper part of the spine. Vertebral fractures can be dangerous, since the projection of the brain in the spinal canal, the spinal cord, passes through the spinal canal. Vertebral fractures are particularly dangerous when the fracture causes the spinal canal to narrow, thus potentially compressing the spinal cord and all its neural pathways. If the vertebrae are severely displaced, the spinal cord may be damaged so badly that the patient will be paralysed. Depending on the location of the fracture and the severity of the damage done to the spinal cord, the patient may lose mobility in the legs or in both the legs and the arms. Classic vertebral fractures are those caused by diving into overly shallow water, which will cause the head and neck to bend. If the bending is severe, this may cause fractures to the upper cervical vertebrae. If the fracture causes major damage to the spinal cord in the neck, the patient may suffer paralysis of the arms and legs. Falls from great height may result in fractures to the middle and lower parts of the spine. If a person falls from a great height and lands on the feet, the forces will be distributed across the body in such a way that the middle and lower parts have the largest impact and the highest risk to fracture.

© Springer Nature Singapore Pte Ltd. 2017
J. Hendrikse, *This is Our Brain*,
DOI 10.1007/978-981-10-4148-8_21

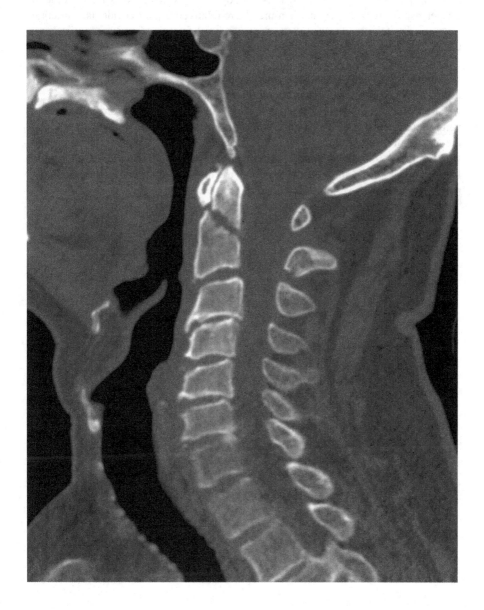

In addition to vertebral fractures caused by accidents, vertebrae may sometimes fracture spontaneously. Generally, such spontaneous vertebral fractures occur in elderly people because their bones have become less solid due to reduced calcium levels in the bones. This is known as 'osteoporosis'. Osteoporosis puts people at greater risk of fractures, even when they have relatively minor accidents. Vertebral fractures due to osteoporosis often involve several fractures in adjacent vertebrae. Generally, these are fractures to the lower part of the spine. Vertebral fractures in elderly people suffering from osteoporosis can be recognised on X-ray images, CT scans and MRI scans since these fractured vertebrae are shaped like wedges or triangles whose fronts are much flatter than their backs.

The CT scan on the next page shows an accident-related fracture in a cervical vertebra (circle). Cervical vertebrae should form a neat stack. Both the fronts and the backs of the vertebrae should be perfectly aligned. In order to determine whether the vertebrae are properly aligned, try drawing a straight line along the front and back of the spine. If the vertebrae are unevenly stacked, the displacement may be caused by a vertebral fracture.

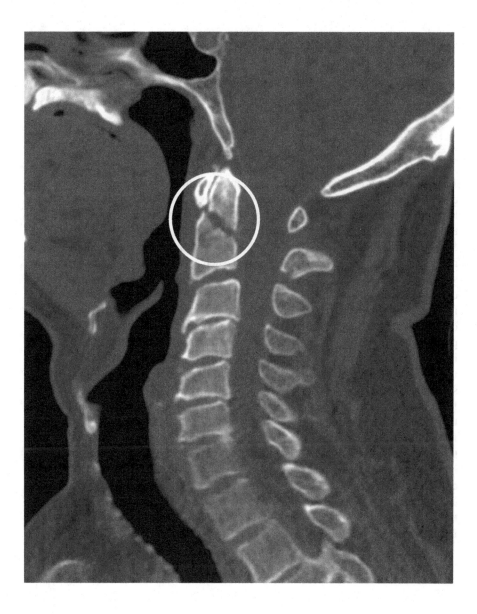

Chapter 22
Encephalitis

The MRI scan on the next page shows an inflammation of the brain, also known as encephalitis. Inflammation of the brain is a relatively rare phenomenon. It is much less common, for instance, than cerebral infarctions. However, due to an increase in the number of patients with reduced immunity, the number of people suffering encephalitis has increased in recent decades, as encephalitis is much more common in persons with a weakened immune system, such as AIDS patients, whose immune system is under attack from the HIV virus. In addition, there are medical procedures, such as organ transplants, which involve a deliberate reduction of the body's immune system so as to ensure that the body does not reject the newly transplanted organ. There are other, rarer factors which cause encephalitis, as well. For instance, patients with endocarditis (inflammation of the heart valve) are at risk of this inflammation spreading to the brain through their blood. Furthermore, severe inflammations of the air-filled cavities behind the ear or in the frontal sinuses may sometimes break through the skull, thus causing the brain to get inflamed, too. Inflammation of the brain can be found in several places, the most common locations being the protective membranes covering the brain and the brain tissue itself. Inflammation of the protective membranes, also known as meningitis, can be hard to detect on a CT or MRI scan. The method used to determine whether a patient has meningitis is a lumbar puncture, followed by analysis of the cerebrospinal fluid thus collected. To this end, a little cerebrospinal fluid will be collected from the bottom of the spinal canal. Even though the bottom of the spine is far away from the brain, the cerebrospinal fluid in the spinal canal is directly connected to the cerebrospinal fluid around the brain.

© Springer Nature Singapore Pte Ltd. 2017
J. Hendrikse, *This is Our Brain*,
DOI 10.1007/978-981-10-4148-8_22

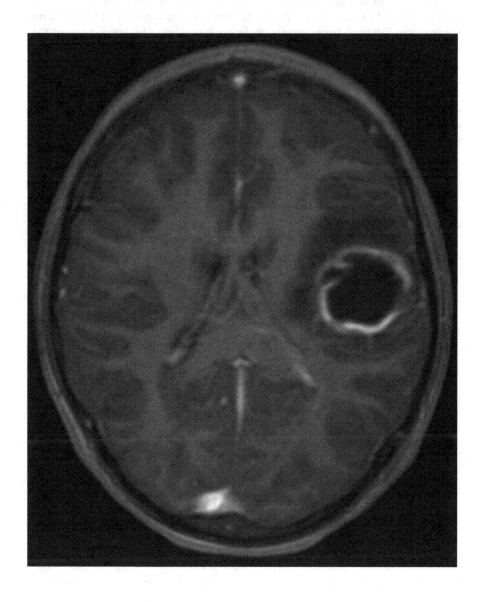

It is easier to collect cerebrospinal fluid from the bottom part of the spinal canal than from the top, which is why punctures are generally performed here. The cerebrospinal fluid collected by means of the lumbar puncture is then examined for the presence of inflammatory cells, which are an indication that the patient has encephalitis or meningitis. If the brain itself is inflamed, this can often be seen on an MRI scan. The MRI scan on the next page shows an inflammation within the brain (circle). The inflamed tissue shows up as a white edge. The entire edge of the inflammation forms a circle. Within this white circle, a dark grey to black area can be observed in the middle. In order to make inflammations more readily visible on CT and MRI scans, patients are often intravenously administered a fluid contrast medium through a drip attached to a blood vessel in the elbow fold. This contrast medium will then spread across all the blood vessels in the body. Normally, the blood vessels in the brain are so well insulated that contrast medium does not leak from them. If there is any inflammation, the blood vessels' level of insulation is sharply diminished, as a result of which the injected contrast medium will leak from the blood vessels. White circles such as the one visible on this MRI scan will be caused by such a contrast medium leak. The centre of the circle is dark because the waste products of the inflammation are located here. The combination of the white edge and the collection of infected material at its centre is called a 'brain abscess'. Treatment of encephalitis often involves a combination of antibiotics and when an abscess is present an operation carried out by a neurosurgeon.

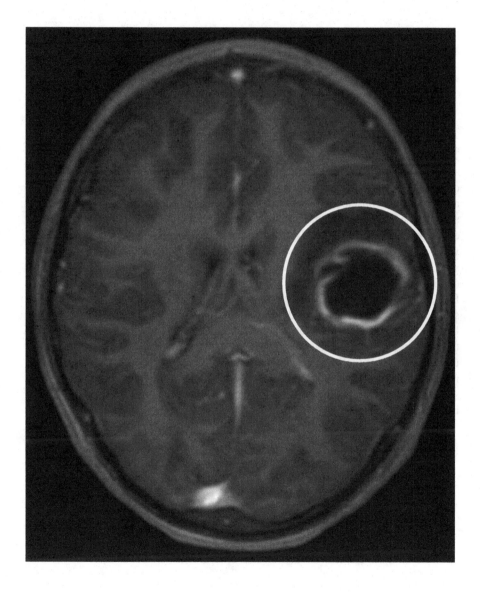

Chapter 23
The Causes of Epilepsy

The MRI scan on the next page shows a small abnormal area in the brain. This abnormal area is causing the patient to suffer from epilepsy. There are various types of epilepsy, as well as several possible causes. Epileptic seizures come with an altered state of consciousness and often involve muscle spasms as well. Sometimes an MRI scan will show damage in the brain which will explain the patient's epilepsy or epileptic seizure. This damage may have been present when the patient was born, or alternatively, it may have developed at a later age. Sometimes, an MRI scan will not be able to show the cause of the epilepsy. Electroencephalography (EEG) may record the abnormal activity of the brain during an epileptic seizure. EEG, in combination with the doctor's knowledge of different types of epileptic seizures, may provide a clue of the location in the brain where the epileptic seizure begins. Young patients sometimes suffer a congenital disorder in the outer layer of the brain, which is also called the 'cortex' or 'grey matter'. Congenital disorders in this outer layer of the brain may cause epilepsy in some patients. If medication is not sufficiently effective, doctors may decide to perform surgery to remove a smaller or larger area of the brain, the objective being the removal of the area of the brain which is causing the seizures.

© Springer Nature Singapore Pte Ltd. 2017
J. Hendrikse, *This is Our Brain*,
DOI 10.1007/978-981-10-4148-8_23

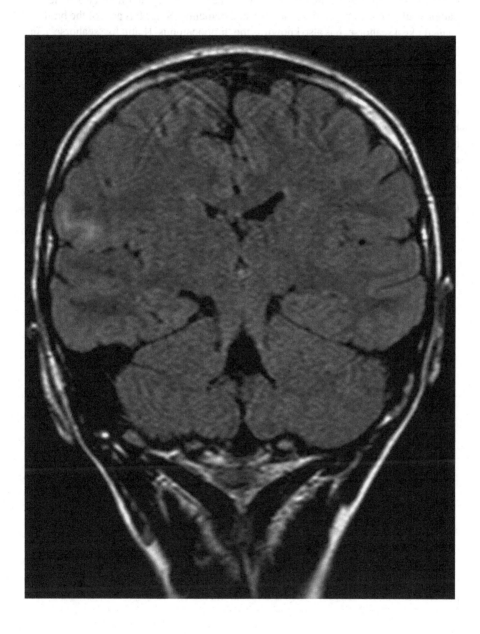

Another area of the brain which may cause epilepsy is hidden deep inside the central brain. This small area is also important to memory and may be smaller in patients suffering from impaired memory and dementia. Since this area of the brain is shaped like a seahorse, it is also known as the 'hippocampus' (Latin for 'seahorse'). In some of the frequently young patients with epilepsy, this seahorse looks abnormal and overly white on MRI scans. In addition, it often looks smaller than one might expect.

Sometimes epilepsy is caused by a brain tumour. Especially, in elderly patients with first time epileptic seizures, the potential underlying cause is a brain tumour or other causes of acquired brain damage. Brain tumours may be growing from the brain tissue itself or be the results of a metastasis (spread) of a tumour situated elsewhere in the body. Both young and old patients who have experienced an epileptic seizure nearly always undergo an MRI scan to determine the cause of the seizure.

The MRI scan on the next page shows a relatively white area of the brain tissue (circle). This relatively white area is a congenital disorder which is causing this patient to suffer epileptic seizures.

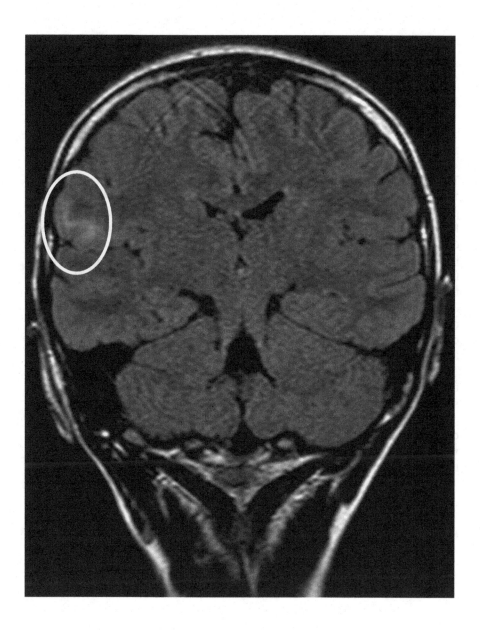

Chapter 24
Non-aggressive Brain Tumours

The MRI scan on the next page shows a non-aggressive brain tumour. 'Non-aggressive' here means that the tumour grows very slowly. As a result, patients with this type of brain tumour can go on living for years, even decades. However, if the tumour starts growing or causing symptoms such as epilepsy, doctors may decide to treat it. Such treatment may consist of an operation, which may or may not be combined with radiation. The nasty thing about brain tumours is that they have thin projections, like an octopus' arms, which protrude into parts of the brain which look normal on the MRI scan. Because of these thin projections, even non-aggressive brain tumours are hard to cure completely. Even years after an operation, a brain tumour may recur. Generally, the tumour will return on the edge of the area that was operated upon. Occasionally, however, the tumour will return at a place far removed from the original site.

There are various types of brain tumours, which are categorised according to the type of cells which make up the tumour. Often a small hole will be drilled into the skull, after which tissue will be removed from the tumour through a thin tube. The cells, which can be observed under a microscope, and their DNA are examined to determine the type of tumour and its level of aggression. The likelihood that a tumour will recur partially depends on what type of tumour it is. Sometimes, mildly aggressive tumours will grow into highly aggressive tumours over the course of time.

© Springer Nature Singapore Pte Ltd. 2017
J. Hendrikse, *This is Our Brain*,
DOI 10.1007/978-981-10-4148-8_24

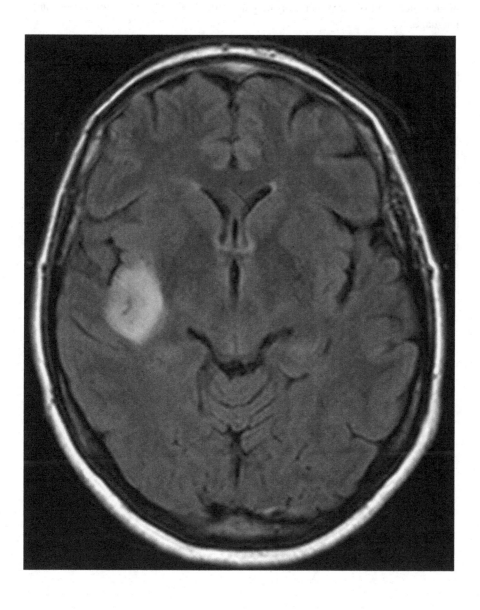

It is not always possible to surgically remove the entire tumour visible on MRI scans, as the tumour may be situated in a vital part of the brain, e.g. parts of the brain which control movement, or the part of the brain which controls speech and language.

The white bulb-shaped area on the MRI scan on the next page is a brain tumour (circle). The surrounding normal appearing brain tissue shows up grey on this MRI scan. It is known that even such normal-looking grey parts of the brain often contain small and thin projections of the tumour.

The distinction between non-aggressive and aggressive brain tumours is made using a microscope, by a specialised doctor called pathologist, who examines the brain tissue removed during biopsy or neurosurgery. By looking at the number of dividing cells and the cells' DNA, doctors can determine how aggressive a tumour is. The more dividing cells, the more aggressive the tumour. MRI scans, too, show how aggressive a tumour is. By intravenously injecting a contrast medium into the elbow fold, and subsequently performing an MRI scan, doctors can determine whether the vessels in the tumour are leaky. Normally, the injected contrast medium will not leak from the blood vessels. The more aggressive the tumour, the more significant the leakage from the blood vessels. The non-aggressive brain tumour on the next page does not have any visibly leaky blood vessels on the MRI scan performed.

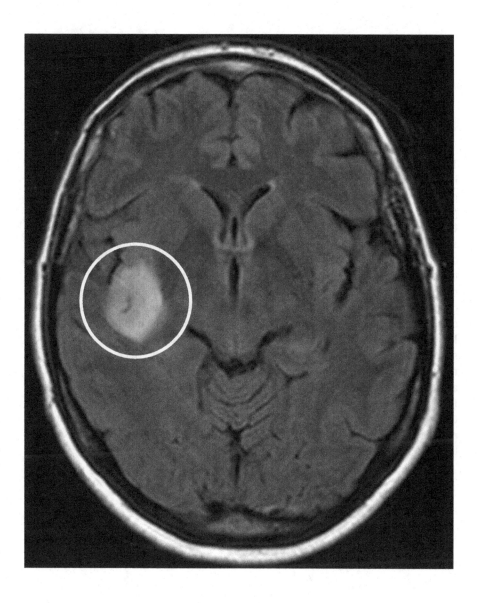

Chapter 25
Aggressive Brain Tumours

The MRI scan on the next page shows an aggressive brain tumour. The thin and invisible projections of an aggressive brain tumour constitute a very serious problem. With an aggressive brain tumour, the octopus' arms are very long, and at present, it is next to impossible to completely cure an aggressive brain tumour. Unfortunately, most patients die within two years of being diagnosed with an aggressive brain tumour, as tiny projections will live in many more places than we can see on an MRI scan. Therefore, MRI scans often result in the scale of the tumour's spread being underestimated. Brain surgery is dangerous. If a surgeon removes too much of the brain, this may result in the patient becoming severely disabled. Therefore, the advantages of a large operation must be offset against the disadvantages of a severe disability. Since small traces of the tumour are likely to be left at the edges of the surgical cavity and probably farther from the tumour, too, supplementary treatment is required in addition to surgery. Such additional treatment consists of radiation (radiotherapy) and medication (chemotherapy). The addition of radiotherapy and chemotherapy has helped us make progress in the last few years. Patients with aggressive brain tumours now have a longer life expectancy than they used to. But even after the triple treatment strategy of surgery, radiation and medication, brain tumours will often recur within a few years of the therapy.

© Springer Nature Singapore Pte Ltd. 2017
J. Hendrikse, *This is Our Brain*,
DOI 10.1007/978-981-10-4148-8_25

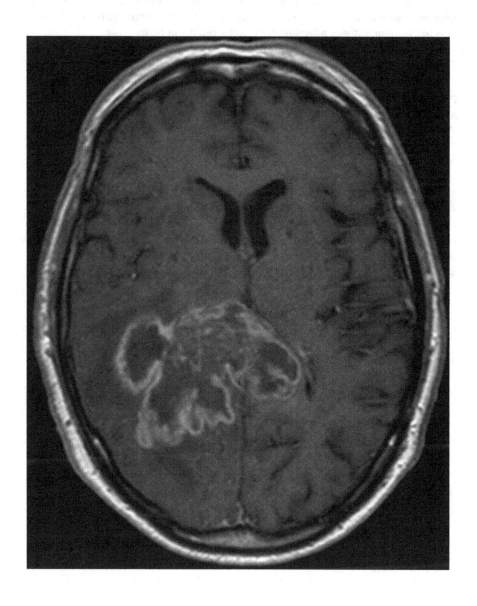

If a tumour does return, treatment options are even more limited. Increasingly, patients take medication and undergo for a second-time surgery or radiation with recurrence of brain tumours. With aggressive brain tumours, doctors are placing their faith in medications and other treatments which eliminate tumour cells in a targeted fashion without affecting the healthy brain too badly.

The white area on the image on the next page is an aggressive brain tumour (circle). This brain tumour has leaky blood vessels, thus causing the injected fluid (contrast medium) to drip from the vessels inside the tumour. Before such an MRI scan can be performed, a contrast medium must be injected into the patient's blood vessels. This is generally done by intravenously administering the contrast medium through a drip attached to the patient's elbow fold, from where it will spread across the patient's blood vessels. Contrast medium will not leak from healthy blood vessels inside the brain. Nor will it leak from a non-aggressive brain tumour. Therefore, when contrast medium leaks inside a brain tumour, there are good grounds to assume that the tumour is aggressive. In many cases, a small hole will be drilled into the skull, after which tissue will be removed from the tumour through a thin tube to prove that the tumour is aggressive based on a high number of dividing cells. Sometimes, this biopsy is combined with a more extensive removal of the brain tumour in one neurosurgical operation. The definitive diagnosis of an aggressive brain tumour will always be made using a microscope, by determining the number of dividing cells and analysing the DNA.

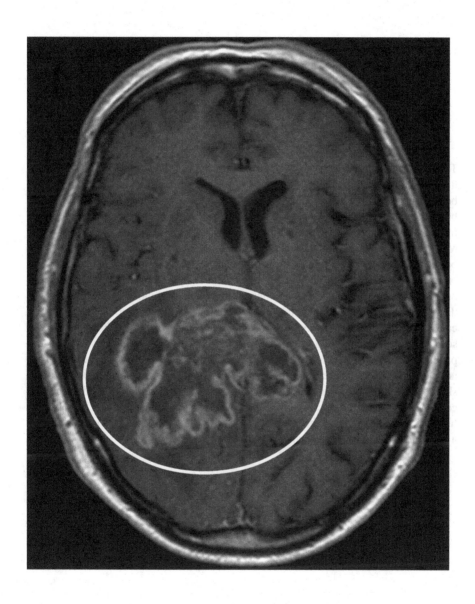

Chapter 26
Metastatic Brain Tumours

The MRI scan on the next page shows a brain tumour. This tumour spread from a tumour elsewhere in the body. A tumour which has spread from another place in the body is called a 'metastasis'. Since this tumour is located inside the brain, it is a brain metastasis. Nearly all types of tumours can give rise to brain metastases. For instance, lung cancer, breast cancer and skin cancer can all metastasise to the brain. Most patients with brain metastasis are known with a tumour elsewhere in the body before being diagnosed with a brain metastasis. However, in some patients, the brain metastasis will be the first tumour to cause symptoms, without the patient being aware that he or she has another tumour. Once a brain metastasis has been discovered, doctors will immediately start looking for the tumour elsewhere which caused the metastasis. This may involve imaging of the rest of the body with a CT scan. In some patients, it is very hard to determine from an MRI scan whether the patient's brain tumour is a tumour from the brain tissue itself or a metastasis from another tumour elsewhere in the body. Once a small hole has been drilled into the skull and a bit of tissue has been removed from the tumour through a thin tube, the cells examined under a microscope will show the cause of the tumour. The cells of a tumour which grows from the brain tissue itself look very different from the cells of a metastasis of a tumour elsewhere in the body.

© Springer Nature Singapore Pte Ltd. 2017
J. Hendrikse, *This is Our Brain*,
DOI 10.1007/978-981-10-4148-8_26

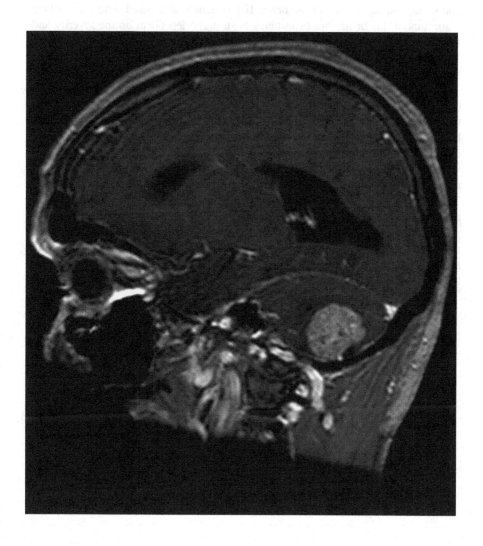

When a metastatic brain tumour is found on an MRI scan, it is vital that doctors check very carefully whether they are dealing with just the one metastasis or with multiple metastases. Sometimes, a tiny second or third metastasis will be present. Needless to say, it is crucial to the treatment to be administered that doctors know whether there are any other metastases present. These days, it is possible to treat one or several metastases in the brain. If a patient is diagnosed with a metastatic brain tumour, his or her life expectancy is strongly dependent on the severity and size of the original tumour and of the metastases in other parts of the body. Tumours from elsewhere may metastasise into the bones of the skull as well as into the brain tissue itself.

The white circle on the MRI scan on the next page is a metastatic brain tumour (circle). This metastatic brain tumour is situated in the cerebellum, where metastases like to go. The MRI scan on the next page was performed after the patient had been intravenously administered a contrast medium through a drip attached to the elbow fold. The contrast medium spread across the body, and since the blood vessels in a metastatic brain tumour leak, the contrast medium then dripped into the metastasis. Contrast medium shows up as an entirely white area on an MRI scan. Tumours growing from the brain tissue itself also often involve leaking contrast medium. Metastases and tumours grown from the brain tissue itself differ in the presence of projections. Unlike tumours growing from the brain tissue itself, metastases do not have these thin projections. Thanks to the absence of these thin projections, metastatic brain tumours often respond well to surgery or radiation treatments.

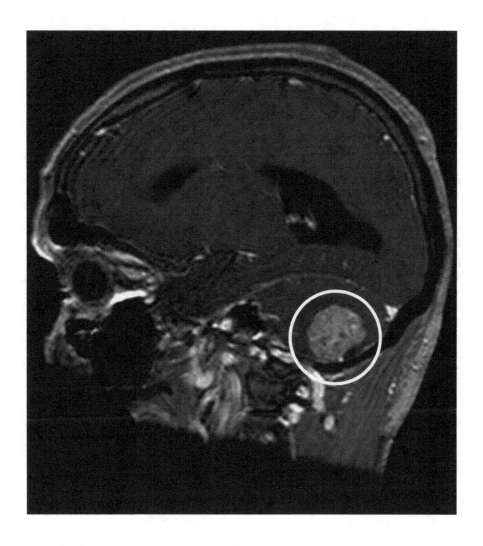

Chapter 27
Metastatic Spinal Tumours

The MRI scan on the next page shows a spinal tumour. Spinal tumours are in the majority of patients caused by metastases from tumours elsewhere in the body. Many tumours are capable of metastasising to the vertebrae. Tumours which often metastasise to the vertebrae are prostate cancer and breast cancer. A metastasis in a vertebra will cause the vertebra to grow weaker, which in turn may cause the vertebra to fracture. The fracture will cause the vertebra to grow flatter. Generally, metastases in the vertebrae will cause the patient to suffer pain. In addition to fracturing vertebrae, metastatic spinal tumours may also grow, thus causing the patient's spinal canal to grow narrower. In the event of a severe narrowing of the spinal canal, the patient's spinal cord and nerve roots may get severely compressed, which may cause the patient to become paralysed. The spinal canal in the upper part of the spine, i.e. the first nineteen vertebrae from the top, contains the spinal cord. The spinal cord is a projection of the brain inside the spinal canal. The spinal canal from the 20th to the 24th vertebrae only contains nerves. Nerves exit the spinal canal through small openings at either side of the spine between all 24 vertebrae. The nature of the patient's symptoms will depend on the location of the narrowing (stenosis) of the spinal canal.

© Springer Nature Singapore Pte Ltd. 2017
J. Hendrikse, *This is Our Brain*,
DOI 10.1007/978-981-10-4148-8_27

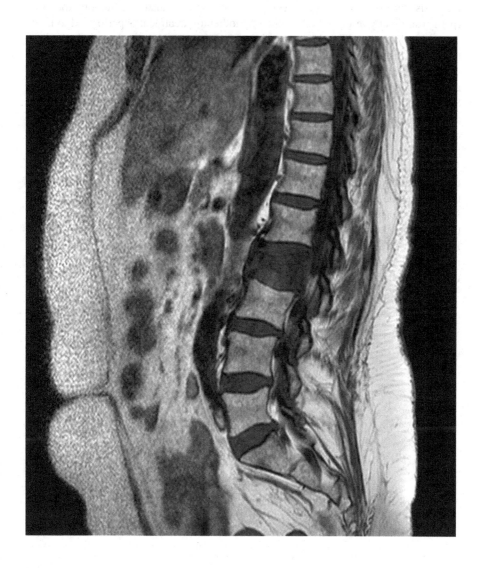

In some patients, metastatic spinal tumours may compress nerves, which may cause the pain and paralysis typically associated with the nerves being compressed. When patients suffer pain, a fracture or a collapsed vertebra due to a metastatic tumour, they are often treated by exposing the affected vertebrae to radiotherapy treatments by the radiotherapy department. If the metastatic tumour is causing paralysis due to a narrowing (stenosis) of the spinal canal, the tumour may be (partly) surgically resected, after which radiotherapy treatment is performed to treat the residual tumour. The doctor administering the radiotherapy is called a radio-therapist. During a radiotherapy session, a CT scanner is used to ensure that the right area (i.e. the metastasis) is receiving radiation. A radiotherapy session involves administering a high dose of radiation to the location of a tumour or metastasis. Since tumours and metastases grow fast, their cells are dividing to produce new cells and tumour growth. Dividing cells are highly sensitive to radiation. The objective of radiotherapy treatment is to eliminate the tumour or metastasis while minimising damage to the healthy parts of the body adjacent to the tumour.

The MRI scan on the next page shows a vertebra with a metastasis, which shows up as a darker shade of grey (circle) than the other vertebrae. The vertebra containing the metastasis has collapsed and is less square than the other vertebrae. Normally, the centre of vertebrae consists of fat tissue called bone marrow. Such fat tissue is why the normal vertebrae in the MRI scan on the next page show up as a lighter shade of grey. In vertebrae containing metastases, the normal fat tissue is replaced with tumour cells, which are more watery and cause the vertebrae to show up darker on this type of MRI scans.

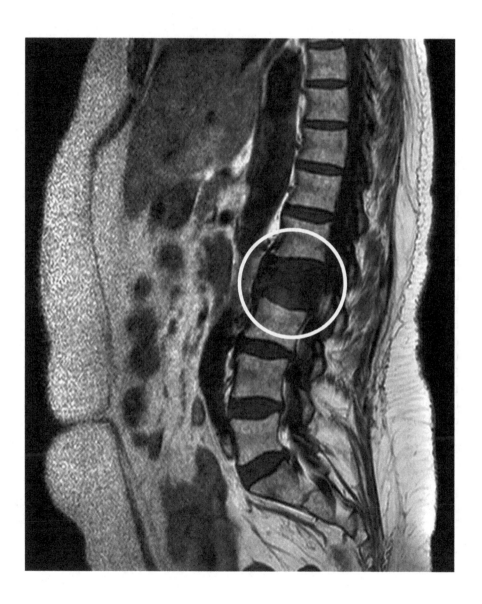

Chapter 28
Too Much Cerebrospinal Fluid

The MRI scan on the next page shows a large amount of cerebrospinal fluid at the centre of the brain. Cerebrospinal fluid can be found at several locations around the brain. Elderly people tend to have more cerebrospinal fluid between the brain and the skull because the brain shrinks, while the volume inside the skull remains the same. A great deal of cerebrospinal fluid can also be found more in the central area of the brain surrounded by brain tissue. This cerebrospinal fluid deep inside the brain is located in several pools which are interconnected. Inside these pools, the cerebrospinal fluid is produced. Since cerebrospinal fluid is continuously being produced, there is a continuous flow of cerebrospinal fluid from one pool to the next. Doctors call these pools 'ventricles'. At some locations, the connections between the various pools are very narrow. If such a connection between two pools of cerebrospinal fluid grows narrower or is completely obstructed, a widening will take place upstream. It can be compared to a dam being constructed inside a river. Since new cerebrospinal fluid is still being produced, the pools of cerebrospinal fluid upstream will grow increasingly wide. In some patients, the cerebrospinal fluid drainage channel will get obstructed without a clear cause. In others, there will be a demonstrable cause.

© Springer Nature Singapore Pte Ltd. 2017
J. Hendrikse, *This is Our Brain*,
DOI 10.1007/978-981-10-4148-8_28

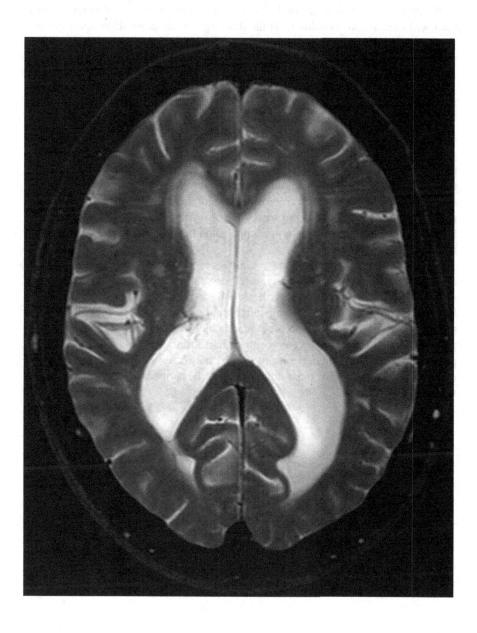

Cerebrospinal fluid drainage may also be impeded by a haemorrhage in the cerebrospinal fluid or an obstructing tumour. Either will result in an increase in size of the pools of cerebrospinal fluid. As we grow older, the pools of cerebrospinal fluid will increase in size due to brain shrinkage. With brain shrinkage over the years, the amount of cerebrospinal fluid on the outside, between the skull and the brain, and also inside the brain, will grow. If the pools of cerebrospinal fluid are found to be greater than expected, we can compare their size with the amount of cerebrospinal fluid between the brain and the skull. If there is more cerebrospinal fluid at both locations, this means brain shrinkage.

If the amount of cerebrospinal fluid in the pools is large, while the amount of cerebrospinal fluid on the outside is normal for the patient's age, the patient probably suffers from a blockage in the cerebrospinal fluid drainage. To treat this problem, a hole can be drilled into the skull. A small tube can then be inserted into the cerebrospinal fluid. A long subcutaneous tube (i.e. a tube inserted under the skin) can then transport the cerebrospinal fluid to the patient's belly. In some patients, it is in fact possible to treat the cause of the overabundance of cerebrospinal fluid, e.g. by creating a new connection between the various pools or by treating a brain tumour.

The MRI scan on the next page shows several severely enlarged pools of cerebrospinal fluid (circle). Normally, cerebrospinal fluid would flow through small openings from these pools to a smaller pool exactly at the heart of the brain. In healthy persons, all pools of cerebrospinal fluid are interconnected, thus allowing the cerebrospinal fluid to flow to other places unobstructed—for instance, from the pools inside the brain to the cerebrospinal fluid inside the spinal canal. Also, the pools of cerebrospinal fluid, centrally located between the brain tissue, are interconnected with the cerebrospinal fluid located at the periphery of the brain between the brain and the skull.

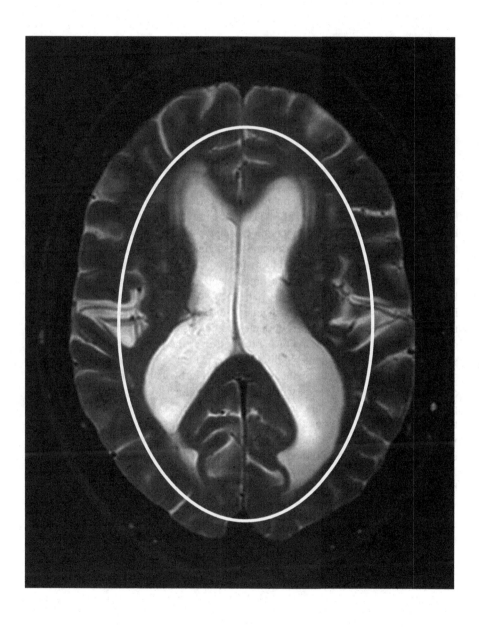

Chapter 29
Dementia and Alzheimer's Disease

The MRI scan on the next page shows a patient whose memory is deteriorating. A deteriorating memory is one of the consequences of dementia. During the early stages of dementia, the patient mainly struggles to remember things which have happened or been said recently. There are several types of dementia, the best-known and most common type of dementia is Alzheimer's disease. In addition to memory loss, dementia impairs many other functions, such as problem-solving skills, learning skills and general information-processing skills. All these symptoms taken together are called 'cognitive impairment'. The patient's symptoms are the key to the diagnosis of dementia or Alzheimer's disease. When a patient's brain is inspected under a microscope, Alzheimer's disease presents with different abnormalities compared to other forms of dementia. For instance, other types of dementia often present with many infarctions spread all over the brain, which is not necessarily the case with Alzheimer's disease. The patient's memory issues and other symptoms may be observed by the patient himself or by their nearest and dearest. In addition, there are many tests which can be performed to determine whether the patient suffers from dementia or Alzheimer's disease. These tests can also be used to record the rate of deterioration over the years. At present there is no medication which is effective against either dementia or Alzheimer's disease. It is possible that exercising and keeping fit in general have a positive effect on the symptoms. However, the symptoms are also affected by many other things which are much harder to control. For instance, moving from familiar surroundings to unfamiliar surroundings will often worsen the symptoms.

© Springer Nature Singapore Pte Ltd. 2017
J. Hendrikse, *This is Our Brain*,
DOI 10.1007/978-981-10-4148-8_29

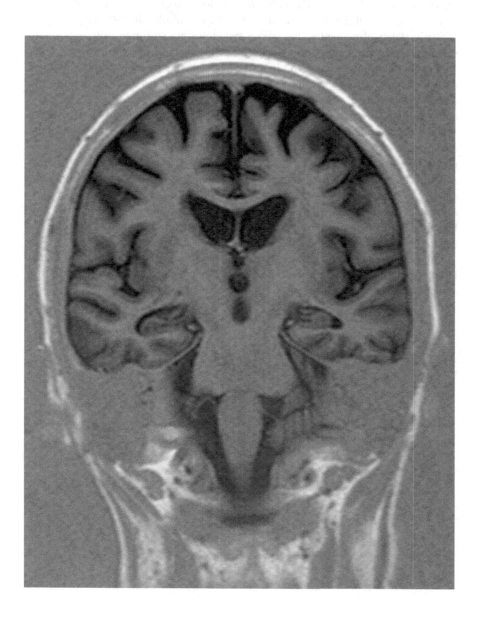

Highly educated people tend to suffer fewer symptoms, or to experience their symptoms at a later age, most likely because such patients are longer able to offset the effects of the disease. MRI scans are used to supplement the information provided by the patient, his or her relatives and the test results. In patients who are in the early stages of dementia, and whose symptoms are still mild, MRI scans may sometimes provide supporting evidence, e.g. by proving that the patient's brain volume has strongly decreased. In some patients, the brain volume will have decreased in certain specific locations, such as the front parts of the brain or the medial lower parts of the brain.

In addition, the part of the brain which looks like a seahorse will often have suffered a reduction in volume. This seahorse is situated deep inside the brain, at the medial base on either side of the brain, and is known to doctors as the 'hippocampus' (small circles). Other changes on MRI scans often found in patients suffering from dementia include many abnormalities commonly found in ordinary elderly people as well, e.g. many white spots, small infarctions and an increased amount of cerebrospinal fluid, both outside the brain and in the pools of cerebrospinal fluid situated at the centre of the brain. In addition, small old microbleeds can be observed in some patients.

On the MRI scan on the next page, the cerebrospinal fluid shows up black. Decreased brain volume can be observed in this patient, as well as an increased amount of cerebrospinal fluid at the front of the brain (large circle).

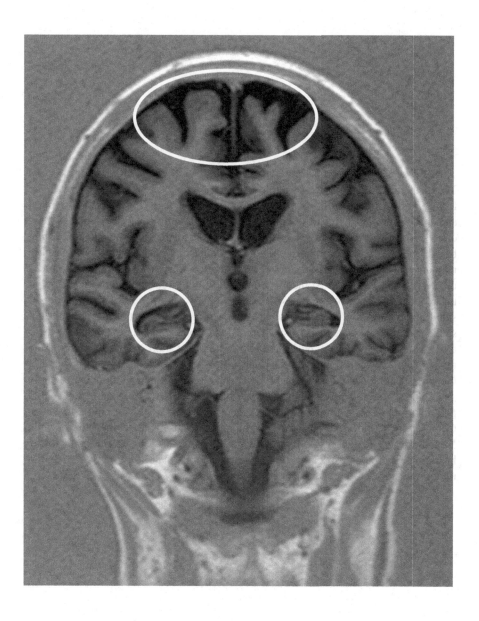

Chapter 30
Arteriovenous Malformation

The MRI scan on the next page shows a pile-up of smaller and larger blood vessels. The main cause of such pile-ups of blood vessels in the brain is a short circuit in the form of a direct connection between the blood vessels carrying blood to the brain and those draining it from the brain. Such short circuits cause the blood to flow directly from the blood vessel carrying the blood to the brain into one carrying blood from the brain. Normally, this blood would have to pass through an entire network of very small blood vessels, called the 'capillaries', where the brain picks up oxygen and nutrients. At the location of the short circuit, the blood will flow to the blood vessels much more easily than at other locations. Since the network of small blood vessels giving off oxygen is bypassed, the site of the short circuit will start sucking blood away from elsewhere. Blood which was originally on its way to other parts of the brain will be sucked to the site of the short circuit. In addition, the total amount of blood flowing to the brain will often increase. Due to the increase in the amount of blood sucked to the blood vessel pile-up, more and more increasingly wide blood vessels will emerge at the site of the short circuit. Doctors call such a pile-up of blood vessels an 'arteriovenous malformation' or 'AVM'. 'Arterio' refers to the arteries (blood vessels carrying blood from the heart to the brain), while 'venous' refers to the veins (blood vessels carrying blood from the brain to the heart). The main risk associated with an arteriovenous malformation is the risk of a rupture of these blood vessels.

© Springer Nature Singapore Pte Ltd. 2017
J. Hendrikse, *This is Our Brain*,
DOI 10.1007/978-981-10-4148-8_30

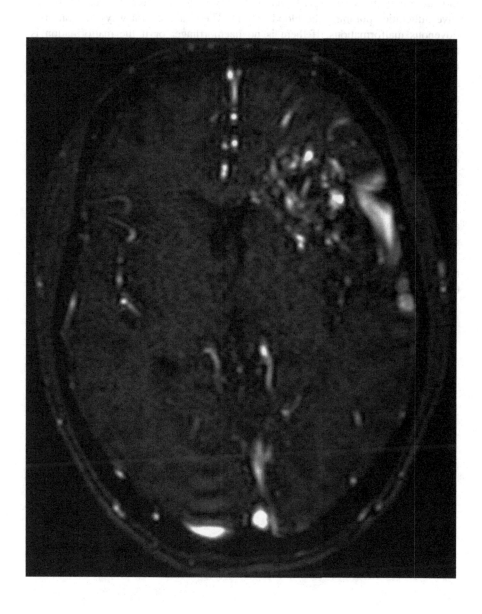

A haemorrhaging ruptured arteriovenous malformation may cause damage to the surrounding brain tissue. Even if there is no haemorrhage, an arteriovenous malformation may grow so large, typically in very young children, that the heart will have difficulties pumping the blood around. There are several ways to treat arteriovenous malformations. If there is no haemorrhage, or if the malformation is incidentally found on an MRI scan, doctors may decide to do nothing for the time being. If the malformation has bled, or if it is growing, the patient has a choice of several treatments. The malformation may be operated on or subjected to radiotherapy, or alternatively, the blood vessels may be closed from the inside. In the latter event, a thin tube will be inserted into the blood vessels from the groin and moved all the way up to the neck through the aorta so as to get to the cerebral blood vessels and the cerebral arteriovenous malformation. Glue may then be injected into the tube to obstruct the blood vessels in the arteriovenous malformation.

It is hard to completely get rid of arteriovenous malformations, and it often takes various different treatments over a longer period of time, such as a combination of surgery, radiotherapy and an intravascular treatment. As mentioned above, waiting and seeing may also be an option, as well, as the various treatments are not without risks including the risk of a cerebral infarction or haemorrhage.

On the MRI scan on the next page, the white lines are all blood vessels carrying blood to and from the brain which are part of an arteriovenous malformation (circle). The vessels crisscross like a plateful of cooked spaghetti.

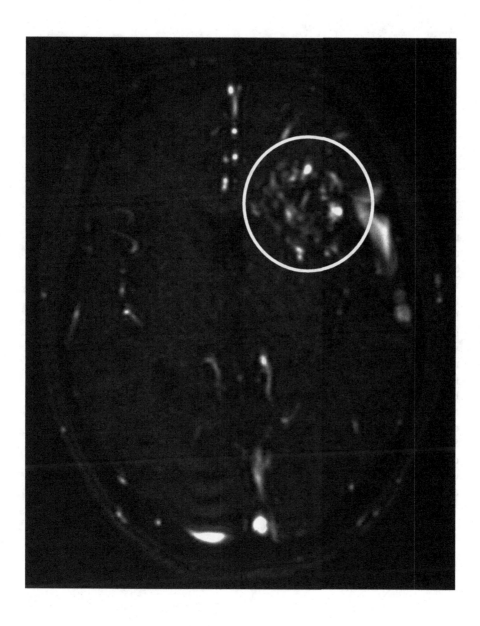

Chapter 31
Tumours Arising from the Meninges

The MRI scan on the next page shows a tumour arising from the protective membrane covering the brain. Doctors call such thin protective membranes 'meninges' and the tumours arising from them 'meningiomas'. Since these tumours are not located in the brain itself, it is much easier to treat them compared to the treatment of brain tumours that are arising from the brain tissue. Generally, MRI and CT scans show a very clear boundary between a meningioma tumour and the brain tissue. In addition, this type of tumour tends to grow very slowly. In these respects, meningiomas are very different from aggressive and non-aggressive brain tissue tumours, which often come with thin projections which are hard to treat. Many elderly people have small meningiomas without realising it. Small meningiomas often present without symptoms. They are sometimes incidentally found when a patient undergoes an MRI scan for a different reason. If such tumours arising from the meninges grow very large (i.e. several centimetres or more in size), they may compress the adjacent parts of the brain, which may cause the compressed brain tissue to get irritated. Irritated parts of the brain may present with an increased amount of fluid signal on an MRI scan. Meningiomas which press on adjacent parts of the brain and so cause the patient to experience symptoms can be operated upon. Since meningiomas are located just outside the brain tissue and have a clearly defined boundary, surgery tends to be successful. After the operation, patients will undergo repeated MRI scans to verify that the tumour has been removed completely and that it is not coming back.

© Springer Nature Singapore Pte Ltd. 2017
J. Hendrikse, *This is Our Brain*,
DOI 10.1007/978-981-10-4148-8_31

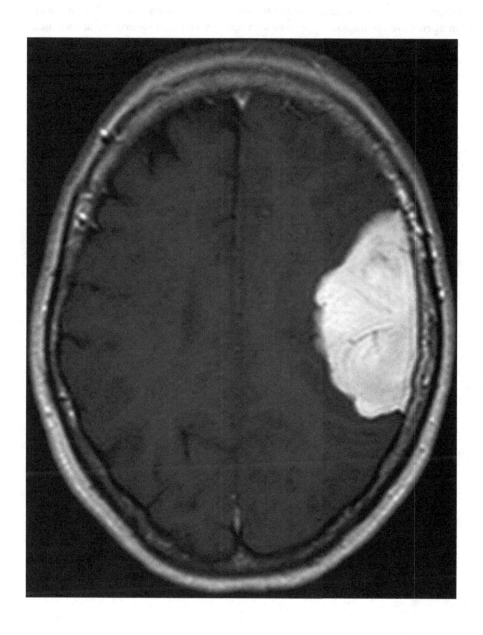

Also more aggressive types of meningiomas exist. Such meningiomas often cause a great deal of irritation to the adjacent parts of the brain. Generally, the border between the meningioma and the brain tissue is much less clearly defined for an aggressive meningioma. Patients with aggressive meningiomas are at increased risk of a return of the tumour in the years after the operation. Some patients with meningiomas have several of these tumours at different locations. If a patient has been diagnosed with a meningioma, it is crucial that the MRI scan is checked carefully for the possible presence of a second, smaller meningioma elsewhere.

On the MRI scan on the next page, a meningioma shows up white (circle). Once a contrast medium has been intravenously administered to the patient's elbow fold, a meningioma will show up bright white on this type of MRI scan. This is because these tumours contain many leaky blood vessels, which will leak the injected contrast medium. Contrast medium dripping from blood vessels shows up bright white on both CT and MRI scans. As you can see from the MRI scan on the next page, there is a clear boundary between the meningioma and the brain tissue next to the tumour.

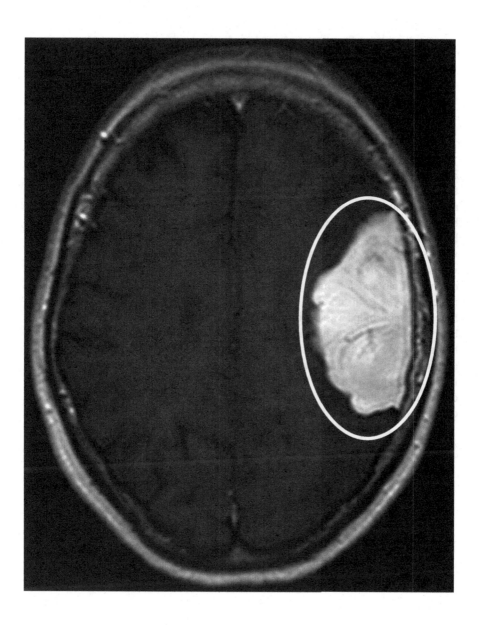

Chapter 32
Co-occurring Multiple Cerebral Infarcts

The MRI scan on the next page shows several cerebral infarcts at different locations. If multiple locations of the brain suffer an infarction, this tells us something about the cause of these infarcts. If the cerebral infarcts hit places which are served by different feeding blood vessels, the cause of the infarcts may be located in the heart. Blood from the heart spreads across the entire body, including all the blood vessels in the neck. If smaller or larger blood clots are propelled from the heart, they will spread all over these blood vessels. Blood clots tend to be made up of old, coagulated blood. Clots may grow in parts of the heart where blood tends to flow more slowly, e.g. in the atrial appendages. Small blood clots from the heart may flow to several blood vessels in the neck, and from there, to different parts of the brain. If such clots get stuck in the various smaller blood vessels deep inside the brain, several parts of the brain may suffer simultaneous infarcts. Since not all clots are the same size, some infarcts may be larger than others. A large clot is more likely to get stuck in a large blood vessel, meaning a large part of the brain will be deprived from blood supply. This will cause a more severe infarction than a small clot which has got stuck in a smaller blood vessel.

© Springer Nature Singapore Pte Ltd. 2017
J. Hendrikse, *This is Our Brain*,
DOI 10.1007/978-981-10-4148-8_32

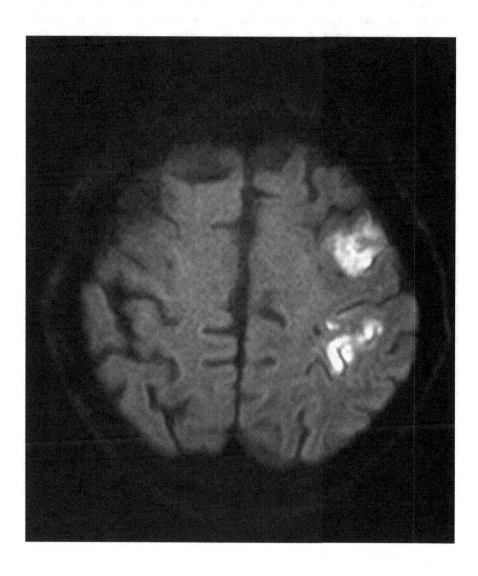

In addition to being caused by coagulated blood in an atrial appendage of the heart, small cerebral infarcts may also occur during blood vessel treatments. Stenoses ('narrowings') of blood vessels of the heart may be treated by inserting small tubes into the groin. From the groin, these tubes are moved upwards via the aorta to the heart region. The movement of such small tubes in blood vessels such as the aorta may cause small clots to break away from the vessel walls. These small clots, too, may cause minor cerebral infarcts. Generally, these will be very small and 'silent' infarcts which do not cause any symptoms to the patient. It is well known that nearly all medical procedures carry a certain amount of inherent risk. The advantages of treating the narrowings of blood vessels in the heart far outweigh the minor risks involved in such procedures. Another cause of multiple cerebral infarcts at different locations are rare conditions such as an inflammation of the walls of the cerebral blood vessels. In most cases, patients will suffer one cerebral infarct at a time. This is generally caused by either a blood clot from the heart, a blood clot from a blood vessel in the neck or a stenosis in a diseased cerebral blood vessel within the skull (atherosclerosis).

The MRI scan on the next page shows two white areas in different parts of the brain (circles). These are cerebral infarcts. A special type of MRI scan has been developed which is highly sensitive to early cerebral infarcts which happened only several hours ago. CT scans often fail to pick up these cerebral infarcts that are only a few hours old. Needless to say, the patient's symptoms are often very obvious even when no clear infarct is visible on a CT scan. CT scans are nowadays often performed in the early phase after an infarct to exclude the presence of blood—a task to which CT scans are eminently suited. Patients can only receive medication to treat a recent cerebral infarct if there is no bleeding. If there is a brain bleeding, these medications may actually exacerbate the haemorrhage. Furthermore, with CT scans performed after the injection of a contrast agent, the location of an occluded cerebral blood vessel can be detected.

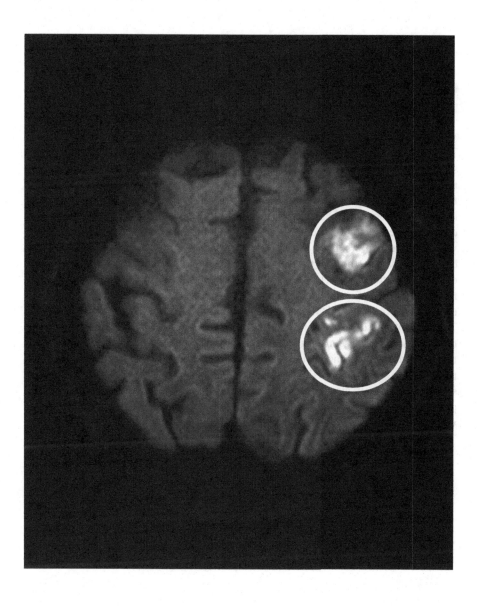

Chapter 33
Vestibular Schwannoma

The MRI scan on the next page shows a small tumour arising from the nerve which serves the vestibular system (for balance). There are several important cranial nerves at the intersection of the cerebellum and the brainstem. These nerves are the facial (7th) nerve and the vestibulocochlear (8th) nerve, which supplies balance and hearing. Doctors call these nerves the 7th and 8th cranial nerves. There are 12 cranial nerves in total on either side of the brainstem. From the 8th nerve, the vestibular part is for balance and the cochlear part is for hearing. These 7th and 8th cranial nerves pass through the same small canal, the internal acoustic meatus. The external acoustic meatus is what you feel when you put in the ear buds of your mobile phone. The internal acoustic meatus is hidden deep inside your skull and can only be seen on CT and MRI scans. A thin protective sheath covers the nerves. Tumours of these protective sheaths are quite common at the nerves passing through the internal acoustic meatus. In most cases, such tumours will arise from the vestibular nerve. Since both the 7th and 8th cranial nerve pass through the same narrow canal tumours arising from the one nerve (also known as vestibular schwannoma) may also cause symptoms arising from the other nerve. Many patients present with severe loss of hearing in one ear. Normally, ageing people with hearing problems will slowly loss hearing in both ears.

© Springer Nature Singapore Pte Ltd. 2017
J. Hendrikse, *This is Our Brain*,
DOI 10.1007/978-981-10-4148-8_33

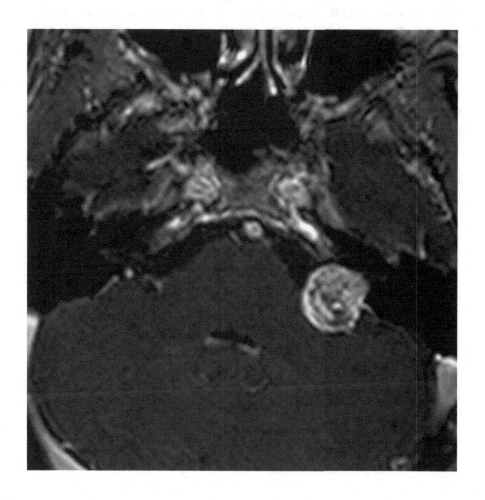

If a patient reports a severe loss of hearing in one ear, doctors may perform an MRI scan to determine whether there is a tumour in the internal acoustic meatus. The great majority of patients who undergo an MRI scan for this reason turn out *not* to have a tumour. Vestibular schwannomas can be treated with radiotherapy or can be removed surgically. In the event of a small vestibular schwannoma, doctors may decide to perform a regular follow-up of the tumour with potential treatment at a later stage when the tumour grows. Generally, patients with a vestibular schwannoma will undergo an MRI scan once every one or two years, to check whether the tumour is growing. If a vestibular schwannoma grows to several centimetres in size, it may compress several nerves, or even compress and displace the brainstem.

The MRI scan on the next page was performed after administering a contrast medium injected through a blood vessel in the patient's elbow fold. The MRI scan shows that the tumour in the internal acoustic meatus is leaking this contrast medium. As a result of the contrast medium leakage, the tumour shows up bright white on this type of MRI scan (circle).

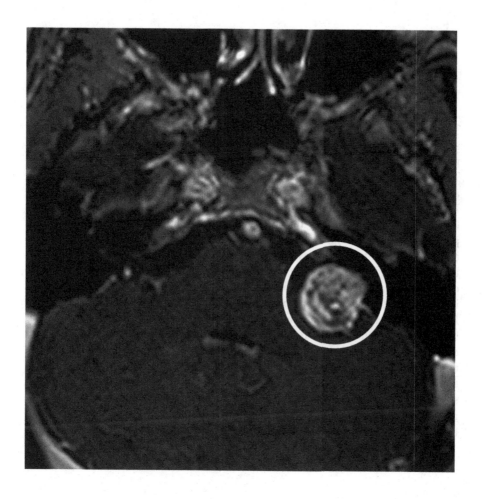

Chapter 34
Pituitary Adenomas

The MRI scan on the next page shows an image at the exact centre of the brain. It shows an image of the brain a top-to-bottom direction which can be imagined when placing your finger on one ear and go to the other ear via the top of your head. There is a protrusion present at the base of the brain. The contents of this protrusion are in healthy subjects vital to the production of several hormones in our body. Doctors call this protrusion the 'pituitary gland'. The release of hormones into our body is largely regulated by this small command centre. Some of the hormones produced by this gland have a direct effect on the body, such as the growth hormone and prolactin. Other hormones secreted by this gland stimulate other parts of the body to secrete hormones themselves. For instance, the thyroid, adrenal glands and ovaries are all regulated by hormones produced by the pituitary gland. Even though it is just a tiny gland within the protrusion, almost 10% of all brain tumours grow from the pituitary gland. Since such tumours (also known as 'adenomas') produce hormones, even small tumours may present with symptoms. Pituitary tumours are called adenomas. These pituitary adenomas are divided into two categories. First, tumours which exceed 1 cm in size which are called macroadenomas. Second, tumours smaller than 1 cm in size which are called microadenomas. If a doctor suspects that a patient suffers from a pituitary adenoma, the patient may be asked to undergo an MRI scan. After contrast medium injection, via a vein at the elbow fold, the MRI scan will be very likely to pick up small pituitary tumours.

© Springer Nature Singapore Pte Ltd. 2017
J. Hendrikse, *This is Our Brain*,
DOI 10.1007/978-981-10-4148-8_34

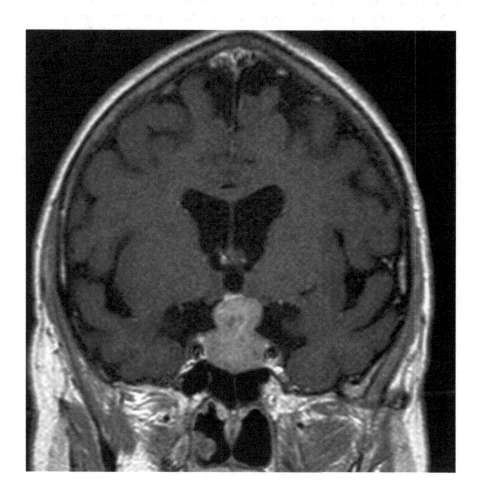

Larger pituitary adenomas can be identified on MRI scans even without the injection of contrast medium. If the tumours grow larger, they may expand in various directions. If a tumour grows upwards, it may compress the optic nerve. If the pressure on the optic nerve becomes significant, some of the patient's visual field may be obscured, thus resulting in his or her visual field to be narrowed. It is as if the patient is wearing blinkers, limiting the outer edges of his or her visual field.

What type of treatment a patient will receive depends on the type and size of his or her pituitary tumour. Pituitary adenomas can be divided into several categories, depending on the type of hormones produced by the tumour. Some adenomas are inactive, which is to say that they do not secrete any hormones. Such adenomas can be surgically removed or treated with medication or radiotherapy. In the event of a tumour pressing on the optic nerve, it is crucial that the tumour is surgically removed, so as to ensure that the optic nerve is no longer compressed. If a tumour returns following treatment, radiotherapy is an option to treat the tumour or a second operation. Nearly 15% of all elderly people have a small pituitary adenoma without experiencing any symptoms. If such small adenomas are incidentally found on MRI scan made for another reason, they are generally left untreated. In these patients, the MRI scan can be repeated after one or several years to exclude growth of a small adenoma. In addition to tumours, small fluid bubbles may be observed inside the pituitary gland. These fluid bubbles generally do not cause any symptoms and do not require treatments.

The MRI scan on the next page shows a tumour shaped like a snowman (circle). This snowman is a pituitary tumour whose size exceeds 1 cm, i.e. a macroadenoma.

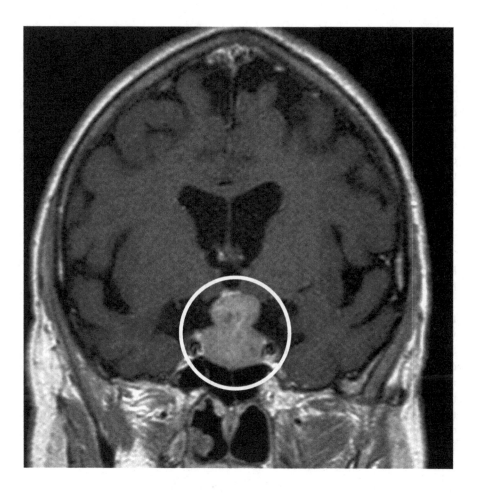

Chapter 35
Otitis: Inflammation of the Ear

The image on the next page is a CT scan of the cavities deep inside the ears on either side of the head. Hearing involves the conversion of a vibration in the air into an electrical stimulus which enters the brain via the nerves. The external acoustic meatus starts at the place where you put an ear plug into your ear. Further down this canal, there is a thin membrane. The external acoustic meatus is on the outside of this membrane and is exposed to the outside air. On the inside of the membrane is a small cavity. This cavity is called the 'middle ear' by ENT doctors (Ear Nose and Throat), while the thin membrane is called the 'tympanic membrane' or 'eardrum'. The middle ear houses three tiny bones which transport sounds further into the ear. These three tiny bones make up a chain which passes sounds on to the inner ear. The inner ear is hidden deep inside the bone of the skull. In this inner part of the bone, the vibrations produced by sounds are converted into electrical stimuli by the hair cells. On a CT scan, the inner ear looks like the spirals seen on a snail's shell. This is the cochlea. The hair cells are situated inside this cochlea. If people listen to loud music too often and too long, the hair cells in the cochlea may get damaged, which may cause hearing loss. Of course, this is not the only reason why people experience hearing loss. For instance, they may suffer from otitis, an inflammation of the ear which may involve the presence of fluid in the middle ear, which is normally filled with air. If there is fluid inside the middle ear, the patient may temporarily experience some hearing loss in one ear. Children sometimes have a tube inserted into the thin membrane, in the external acoustic meatus. This tube causes the air in the middle ear and the air outside to be connected.

© Springer Nature Singapore Pte Ltd. 2017
J. Hendrikse, *This is Our Brain*,
DOI 10.1007/978-981-10-4148-8_35

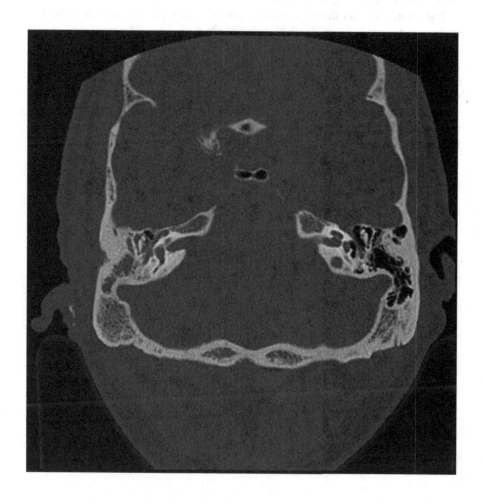

On the CT scan on the next page, the cavities behind the ear show up dark grey on one side and black on the other side. If the left and right sides look different on a scan, this is often the first indication that a problem must be identified. Needless to say, doctors must always consider the possibility that the condition may be affecting both sides. Furthermore, it is important when comparing two sides that doctors always take into account what a particular part of the body—in this case, the middle ear—normally looks like on a CT scan.

The cavities behind the ear normally show up black on CT scans since the air which is supposed to be inside them shows up black on these CT scans, just like the air outside the skull shows up black on CT scans. The fact that the middle ear on the other side shows up greyish is an indication that there is fluid inside the middle ear (circle). The bone surrounding this fluid is bright white because we are dealing here with a particularly strong and thick bone. A few greyish areas can be observed inside the bone. These greyish areas are the cochlea and the thin canals making up the vestibular system (arrows). As said the cochlea is snail's shell structure crucial for hearing and the vestibular system is the structure crucial for balance.

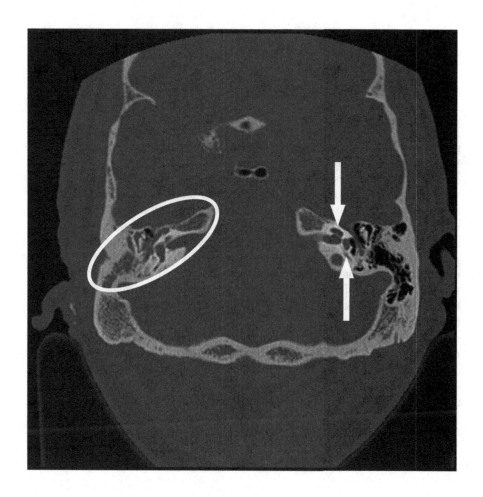

Chapter 36
The Eyes and the Optic Nerve

The CT scan on the next page shows the eye muscles and the optic nerves. This particular CT scan shows a patient with an overactive thyroid, a condition also known as hyperthyroidism. Hyperthyroidism may result in a swelling of the muscles behind the eyes. In many cases, the swelling of the muscles will coincide with an increase in the amount of fat surrounding the eye muscles. Since space is limited behind the eyes, the swelling of the eye muscles will cause congestion, which in turn will push the eyes forward. In severe cases, this forward pressure may affect the patient's ability to move his or her eyes, and the front of the eyes may grow dry because the patient is unable to properly close the eyelids. If a patient suffers swollen eye muscles due to a thyroid issue, the first line of treatment will be to administer proper therapy for the thyroid problem. If the patient suffers badly bulging eyes, and if other treatments have proved ineffective, the patient may have to undergo surgery.

While CT and MRI scans are often performed of the human brain, the eyes do not undergo MRI or CT scanning very often. Generally, ophthalmologists (eye doctors) are able to inspect and diagnose the great majority of eye diseases from the outside. CT and MRI scans of the eye are only performed if there are suspected abnormalities behind the eyes, such as abnormalities of the eye muscles, an inflammation, or a tumour. Tumours behind the eye, such as optic nerve tumours, can be detected on these CT and MRI scans.

© Springer Nature Singapore Pte Ltd. 2017
J. Hendrikse, *This is Our Brain*,
DOI 10.1007/978-981-10-4148-8_36

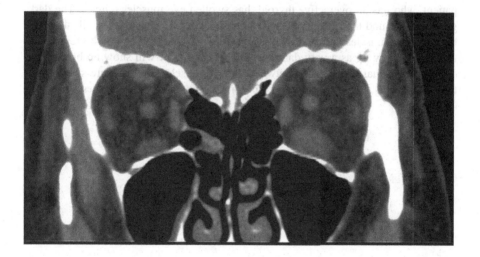

CT scans sometimes show fractured bones around the eyes in patients who have been in an accident. Occasionally, a CT scan will be used to check whether there are any wooden or metal splinters in the eye. Obviously, the eyes are often included in regular MRI scans of the brain, or of the nasal cavities. Doctors looking at such scans will often observe incidental findings regarding the eyes which are not significant to the patient. For instance, they will observe intraocular lenses in patients who have undergone cataract surgery. In addition, they will observe that some people with particularly strong glasses have egg-shaped eyes. Instead of round eyes, these people will have oval eyes. Eye-related abnormalities may also be an indication that the patient suffers a particular disease. For instance, optic neuritis (an inflammation of the optic nerve) may be a first sign of multiple sclerosis (MS). Furthermore, a tumour or inflammation of the nasal cavity may break through a bone and spread in the direction of the eyes.

On the CT scan on the next page, the eye muscles show up grey (arrows). The patient, who has an overactive thyroid, has swollen eye muscles (circle). The dark grey areas around the eye muscles are fat tissue. On CT scans, bones will show up bright white, muscles and brain will show up grey, water will be dark greyish, fat tissue will show up as a very dark shade of grey, and air will show up black. The optic nerve (square) is situated between the eye muscles. With severe swelling of the eye muscles, the optic nerve may become compressed between these muscles.

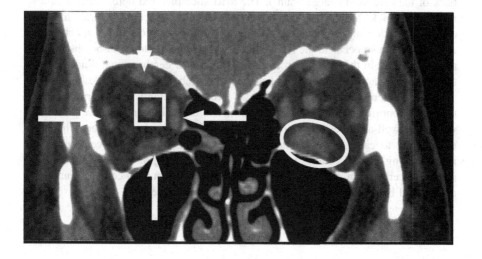

Chapter 37
The Cerebellum

The MRI scan on the next page shows the cerebellum ('little brain') at the lower back of the cerebrum ('large brain'). The MRI slice pictured here, which is a few millimetres thick, shows the exact centre of the brain, dissecting the brain into two equally sized left and right halves. Many of the medical conditions affecting the cerebrum can also occur in the cerebellum. Regular ageing, too, can be observed in the cerebellum, in the form of an increased amount of cerebrospinal fluid between the brain folds and increased fluid between the cerebellum and the skull.

Like the cerebrum, the cerebellum is prone to infarcts, haemorrhages and tumours. Abnormalities such as infarction in the cerebellum may cause positional vertigo, as well as quite a few other symptoms. There are also disorders which are more likely to cause abnormalities in the cerebellum than in the cerebrum. For instance, long-term alcoholics may experience a shrinking of the cerebellum. As a result, the cerebellum will grow even smaller, and the amount of cerebrospinal fluid around and between the folds of the cerebellum will increase.

© Springer Nature Singapore Pte Ltd. 2017
J. Hendrikse, *This is Our Brain*,
DOI 10.1007/978-981-10-4148-8_37

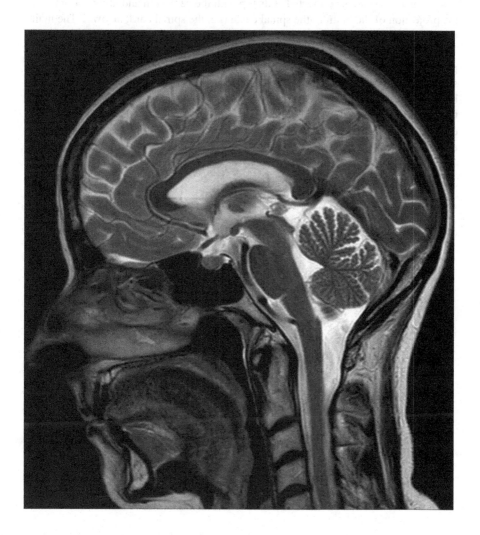

The MRI scan on the next page shows a small pool of cerebrospinal fluid at the front of the cerebellum, coloured white (small circle). This pool of cerebrospinal fluid is situated between the cerebellum and the brainstem. Cerebrospinal fluid flows to this pool through a thin canal from the larger pools centrally located in the cerebrum. From this small pool next to the cerebellum, the cerebrospinal fluid flows either to the spinal canal or to the cerebrospinal fluid between the brain and the skull. The brainstem is connected at its top with the cerebrum and at its bottom with the projection of the brain in the spinal canal (i.e. the spinal cord; arrows). The main blood vessels supplying blood to the cerebellum are at the front of the brainstem. Through various side branches, these blood vessels ensure that the cerebellum is perfused with blood to supply oxygen and nutrients. The cerebellum is situated in a small and relatively closed-off area (circle). It borders a bone of the skull at the back. There is a strong membrane at the top which separates the cerebellum from the cerebrum. Since the cerebellum is located in such a small space, a medical condition such as a haemorrhage or a tumour may cause the cerebellum to get compressed. In some patients, neurosurgery is needed to treat such a compression.

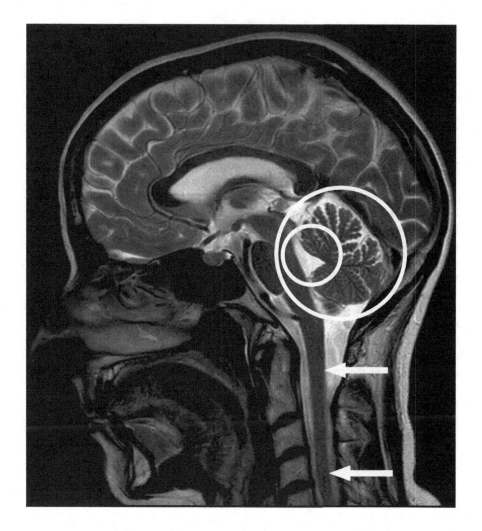

Chapter 38
The Brain, Age 0 to Age 100

The MRI scan on the next page shows a newborn baby's brain. Newborns' brains look completely different from adult brains. They mainly differ in the degree of 'myelination' (maturisation) of white matter. The neural pathways in the white matter myelinate in the first two years of life. What this means is that a fatty layer (the so-called myelin sheath) will grow and cover the neural pathways. This fatty layer could be considered a form of insulation. Without this fatty layer, the brain looks rather watery on an MRI scan. The formation of this fatty layer around the neural pathways follows a set pattern. The neural pathways at the centre of the brain will be the first to grow a fatty layer. Over the first two years of life, this fatty layer will spread to the outside of the brain. The neural pathways which provide motor innervation (the motor cortex) already have a fatty layer when we are born.

Thanks to this insulating layer, signals can be sent from one point to the next much faster. Brain diseases in newborns may affect the myelination process. As an example, the myelination process may be interrupted, with slowing down of the formation of the fatty layer covering the neural pathways or diseases having a negative affect on the composition of the insulating layer. Alternatively, previously created insulating layers around the neural pathways may be broken down. If the formation of the insulating layers is somehow interrupted, an infant's development may be arrested. MRI scans are much better than CT scans at assessing the degree of myelination of the brain.

© Springer Nature Singapore Pte Ltd. 2017
J. Hendrikse, *This is Our Brain*,
DOI 10.1007/978-981-10-4148-8_38

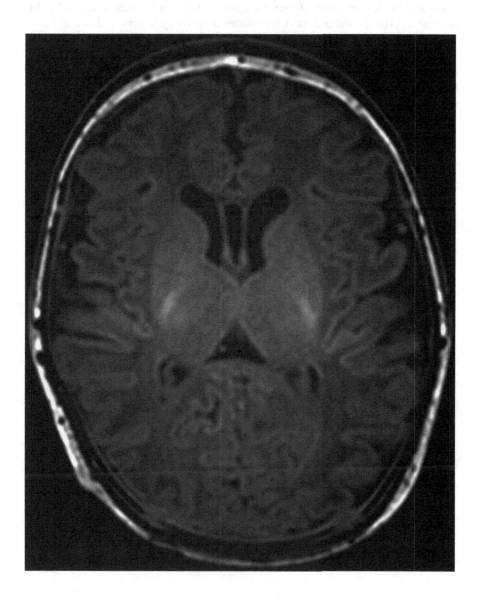

The MRI scan on the next page shows the human brain at age 0. The white areas on the MRI scan are neural pathways which are already covered with an insulating layer (circles). In this particular MRI scan, the outer edge of the brain (grey matter) has a slightly lighter shade of grey than the white matter which is situated right underneath. This is because the white matter is still very watery at this age. On a similar type of MRI scan of an adult brain, white matter looks lighter than grey matter because grey matter is more watery than white matter in adults. White matter is less watery in adults compared to newborns because of the development of the fatty and insulating layer covering the neural pathways in the complete white matter.

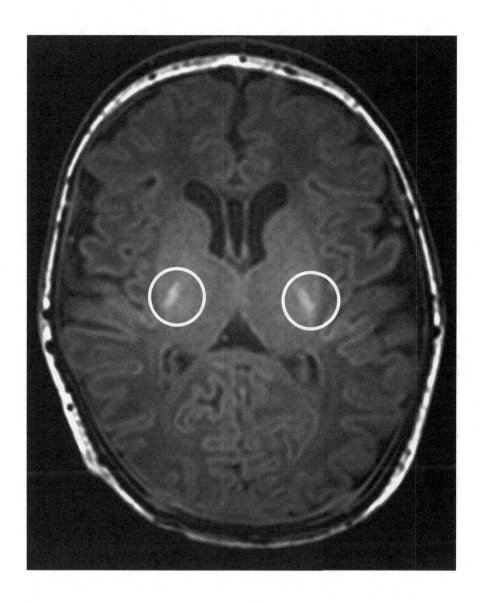

Chapter 39
The Post-operative Brain

The MRI scan on the next page shows the brain following a surgical removal of a brain tumour. Since the tumour was surgically removed, there is now a cavity. Cerebrospinal fluid has filled up this cavity. Scar tissue can often be seen on the edge of such a cavity. The skull, too, will often bear small reminders of the surgery. It goes without saying that the skull must be opened during the surgery and that it must be closed again after the operation. The bone flap created to open a skull will always remain visible to some extent on CT and MRI scans made later in life.

The brain is covered by thin protective membranes. Patients who have undergone surgery will sometimes have a scar and swollen protective membranes at the site of the surgery. These changes to the skull and membranes will not cause any symptoms. Once a patient has had a brain tumour removed, he or she will regularly undergo MRI scans to check whether the tumour was completely removed and to make sure that the tumour can be removed quickly if it returns. Whether or not a tumour recurs is determined by the type of tumour cells and the number of dividing cells which can be seen under a microscope. Many tumours arising from the brain tissue have nasty thin projections which make it hard to remove all the tumour cells. However, brain tumour removal is not the only type of surgery performed on the skull and the brain. For instance, a hole can be drilled into the skull, or a bone flap can be opened, in the event of a severe intracranial haemorrhage. If there is any difficulty in the drainage of the cerebrospinal fluid, a tube can be inserted into the fluid through a small hole in the skull so as to improve cerebrospinal fluid drainage.

© Springer Nature Singapore Pte Ltd. 2017
J. Hendrikse, *This is Our Brain*,
DOI 10.1007/978-981-10-4148-8_39

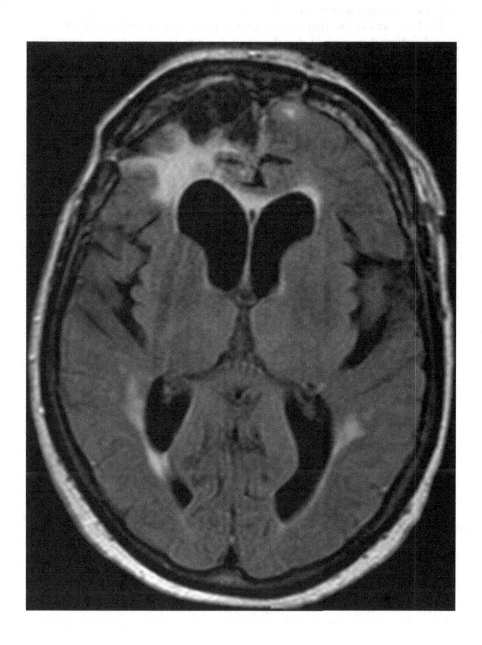

Brain surgery can also be performed to treat severe epilepsy which does not respond to medication. Furthermore, new treatments have recently been introduced, such as a treatment of Parkinson's disease, which involve the stimulating of the brain via thin wires inserted deep inside the brain.

The black area on the MRI scan on the next page is the cerebrospinal fluid which filled the cavity that emerged after the removal of a brain tumour (small circle). The brain tissue directly bordering the edge of the cavity has a slightly brighter shade of white. This white edge is a scar at the border between the surgical cavity and the brain (large circle). A few white spots can be seen a little farther away (arrow). These white spots can be seen on MRI scans of many elderly patients, and these white spots are part of the ageing brain similar to wrinkles in the skin.

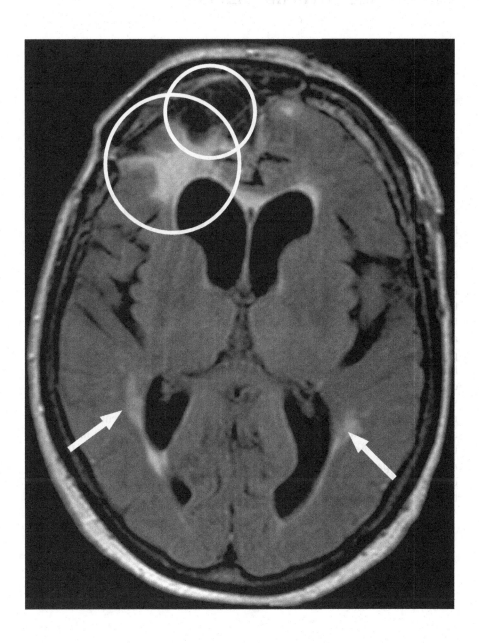

Chapter 40
Many Things Remain Unknown

The MRI scan on the next page looks completely normal, even though the patient was definitely experiencing symptoms. There are many medical conditions which present no, or very few, abnormalities on CT or MRI scans. For an abnormality to show up on a CT or MRI scan, there must be a clear change in the composition of a certain part of the brain. Most visible abnormalities are visible because there has been a change in the amount of water in the brain tissue or because injected contrast medium is leaking from diseased capillaries. Capillaries are the tiniest vessels in the brain which are closest to the brain tissue. Changes in the size of the brain, such as brain shrinkage, must be quite significant to show up in the form of an increased amount of cerebrospinal fluid. A reduction in brain size of just a few percentage points is hard to detect. Computerised calculations and comparisons with other healthy people of a similar age are increasingly enabling doctors to detect small changes related to brain diseases. Severe conditions such as depression or schizophrenia can be difficult to spot on a CT or MRI scan because the changes in the brain are either very subtle or completely invisible. At present, it is hard to attribute subtle changes to particular medical conditions because each individual's brain is unique. As we grow older, our brains all develop certain small abnormalities, comparable to wrinkles in our skin. When assessing a CT or MRI scan, it can be hard to tell signs of normal ageing apart from abnormalities causing patients to have symptoms.

© Springer Nature Singapore Pte Ltd. 2017
J. Hendrikse, *This is Our Brain*,
DOI 10.1007/978-981-10-4148-8_40

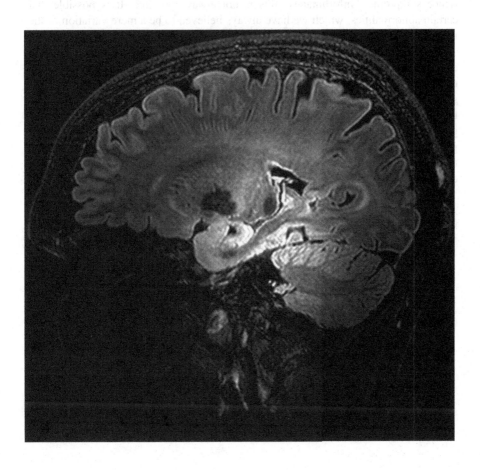

Doctors inspecting MRI scans often come across abnormalities, both subtle and very obvious, whose significance is not entirely clear. These are called 'incidental findings'. Such incidental findings are more likely to be observed on MRI scans of elderly patients. It is vital that doctors be able to distinguish incidental findings, which are unimportant, from abnormalities which actually cause patients to experience symptoms. Unfortunately, this is not always possible. It is possible that certain abnormalities, which we have always believed to be a mere variation of the normal brain, do cause subtle symptoms.

In addition, much research is still needed to improve the therapy for medical conditions such as cerebral infarction, brain tumours, multiple sclerosis and dementia. CT and MRI scans will be able to help us discover new therapies. MRI scans allow doctors to follow small changes over the course of time, thus allowing them to detect the positive effects of, for instance, medication in a small patient population. The efficacy of the more promising medications can then be investigated in much larger patient studies. Such large studies are called 'clinical trials' and these trials cost millions of euros. These clinical trials will remain necessary to demonstrate the efficacy of new medications in reducing patients' symptoms.

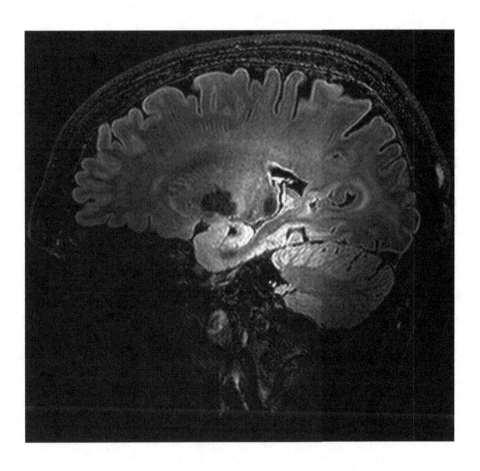

Erratum to: This is Our Brain

Erratum to:
J. Hendrikse, *This is Our Brain,*
DOI 10.1007/978-981-10-4148-8

In the original version of the book, the newly received correction to change the layout has to be incorporated. The erratum book has been updated with the change.

The updated original online version of this book can be found at
DOI 10.1007/978-981-10-4148-8

Afterword
My Own Brain

The image on the next page depicts my own brain. And although I may feel quite young, the first signs of ageing are present. Just like I am getting my first grey hairs and my first wrinkles and pigmented spots on my skin, so my brain is showing subtle signs of change. The amount of cerebrospinal fluid outside my brain and between my brain folds has already increased a little since I was 20, and my brain may already have sustained some subtle damage, judging from the small white spot which can be observed on the MRI scan. However, this is not something I worry about. After all, my brain was able to produce the thoughts and ideas outlined in this book. And although many things can go wrong in the brain, the brain is also eminently capable of repairing itself. For instance, in patients who have experienced cerebral infarction, other parts of the brain will sometimes step up and assume the duties of the part of the brain that has died. Obviously, the brain continues to hold many secrets. Until a few decades ago, we were unable to look inside a living person's skull. However, by reading this book and examining the CT and MRI scans presented in it, you have been able to make a journey through the living brain!

© Springer Nature Singapore Pte Ltd. 2017
J. Hendrikse, *This is Our Brain*,
DOI 10.1007/978-981-10-4148-8

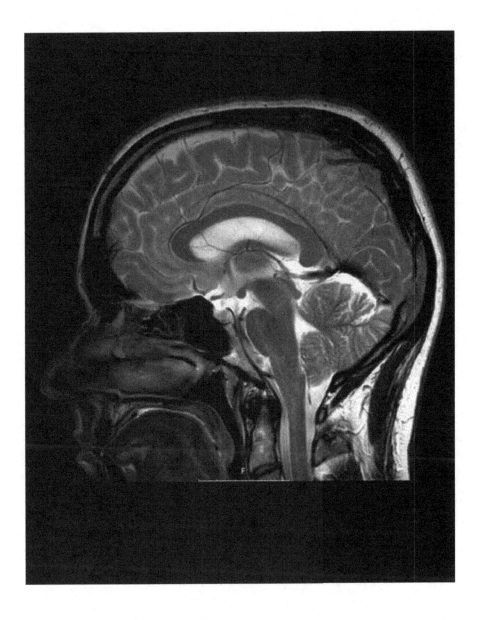

Printed in the United States
By Bookmasters